Creative Music Making

Creative Music Making

William L. Cahn

ROUTLEDGE
NEW YORK AND LONDON

Published in 2005 by
Routledge
Taylor & Francis Group
270 Madison Avenue
New York, NY 10016

Published in Great Britain by
Routledge
Taylor & Francis Group
2 Park Square
Milton Park, Abingdon
Oxon OX14 4RN

Printed in the United States of America on acid-free paper
10 9 8 7 6 5 4 3 2 1

International Standard Book Number-10: 0-415-97281-7 (Hardcover) 0-415-97282-5 (Softcover)
International Standard Book Number-13: 978-0-415-97281-9 (Hardcover) 978-0-415-97281-9 (Softcover)

Library of Congress Cataloging-in-Publication Data

Catalog record is available from the Library of Congress

Taylor & Francis Group
is the Academic Division of T&F Informa plc.

Visit the Taylor & Francis Web site at
http://www.taylorandfrancis.com

and the Routledge Web site at
http://www.routledge-ny.com

CONTENTS

Acknowledgments vii

1. Introduction 1
2. Questions, Questions, Questions! 7
3. A Pedagogy for Musicianship 11
4. Music Making … for Whom? 13
5. The Basics of CMM … Just Do It! 19
6. Improvisation 23
7. Creative Music Making 31
8. CMM Step 1—Playing 35
9. CMM Step 2—Recording 45
10. CMM Step 3—Listening 49
11. CMM Step 4—Questioning 53
12. CMM Sessions at the University of Missouri–Kansas City Conservatory of Music 59
13. Track 4—Duet Improvisation 61
14. Track 5—Trio Improvisation A 65
15. Tracks 6 and 7—Trio Improvisations B and C 71
16. Tracks 8 and 9—Trio Improvisations D and E 75
17. Tracks 10 and 11—Trio Improvisations F and G 81
18. Track 12—Expanding Quartet Improvisation 85
19. CMM in Grade School Classrooms 91

20. Outcomes 97
21. Improvisation in Concerts 99

Building a Homemade Amadinda 103
Examples for Supplementary Listening 105
Recordings of Improvised Music Performed by Bill Cahn 107
Track List for the Enclosed Compact Disc 109
Bibliography 111
Index 113

ACKNOWLEDGMENTS

For the ideas expressed in these pages my gratitude is due the following people:

To John Wyre—for inspiring others through his own example in appreciating the intrinsic beauty of sound, the wonder of intense listening, and the joy of spontaneous music making. He has persistently advocated for the power of intuitive musicianship to be a source of meaningful expression.

To my wife, Ruth—for regularly demonstrating by her example that learning is what education is about, not teaching, and that there is no higher calling than opening the pathways of learning to others.

To my NEXUS colleagues and teachers—Bob Becker, Robin Engelman, Russell Hartenberger, Garry Kvistad, and the late Michael Craden—for years of unrivaled joy in the creation of symphonies of spontaneity.

To Dr. James Snell and all of the participating students at the University of Missouri–Kansas City Conservatory of Music—for wholeheartedly embracing creative music making, for sharing their music on the enclosed CD, and for enthusiastically expressing their thoughts in the workshops at UMKC.

To Dr. Marcia Bornhurst Parkes and her music students at the Ada Cosgrove Middle School in the Spencerport, New York, Central School District—for their participation in photographs contained in this book at the creative music making workshop on June 7, 2004.

To Phil Nimmons—for his playing on the enclosed CD and his participation in photographs contained in this book on November 17, 2003.

To Stephen Adelman, Daniel Barker, Neil Binkley, Linda Forman, Marisa Leva, Julie Roman, and Lenny Seidman—for their participation, shown in photographs contained in this book, at the creative music making workshop on April 24, 2004, at the Painted Bride Arts Center in Philadelphia.

To Dr. Saul Feinberg, my high school music appreciation teacher—for his commitment to the development of good listening skills in all of his students, his love of good music, and his advice in the preparation of this book.

To John Beck, professor of percussion at the Eastman School of Music, and James Campbell, professor of music and director of percussion studies at the University of Kentucky—for their advice and support during the writing of this book.

Bill Cahn

Bloomfield, New York

1

INTRODUCTION

That beauty is not an immediate property of things, that it necessarily involves a relation to the human mind, is a point which seems to be admitted by almost all aesthetic theories … [but] the artistic eye is not a passive eye that receives and registers the impression of things. It is a constructive eye, and it is only by constructive acts that we can discover the beauty of natural things.

—Ernst Cassirer, *An Essay on Man*

MARLBORO, VERMONT, 1968

Improvisation has been at the very core of my own music making through many years of performances with the Toronto-based percussion quintet, NEXUS. Just before NEXUS formed as an ensemble, in the summer of 1968, the five members of the group—Bob Becker, Russell Hartenberger, Robin Engelman, John Wyre, and I—came together at the Marlboro Music Festival in Vermont to perform the percussion parts in Igor Stravinsky's "Les Noces."

John was the timpanist/percussionist in residence there. During his stay he used some of his free time to visit the local antique shops in search of unusual musical instruments, especially bells and gongs. In one shop he found a set of bronze, dome-shaped Japanese temple bells, which he purchased and suspended in the backstage percussion storage room. John had been collecting bells of Asian origin for some time and these new acquisitions were to be added to his collection.

After one of the Stravinsky morning rehearsals, with all of the percussion instruments removed from the stage and arranged back in the storage room, we just started playing around, having fun and experimenting. We amused ourselves by challenging each other to play orchestra excerpts and soon we were playfully distorting them in any imaginative way possible. John's temple bells came into play for a moment in one of the distortions, the fast and tricky xylophone passage from George Gershwin's *Porgy and Bess.* Of course, the temple bell pitches were all wrong because the intonation of the bells fell somewhere in between the notes of the chromatic keyboard scale, all of which made for a great outpouring of fun and laughter.

One result of this shared experience was that our friendship increased and a mutual willingness emerged to go beyond the restrictions with which our musical thinking had been formally educated. Gradually from that time forward in our music making, "notes" and technical issues became diminished in importance and joyful expression became more of a concern.

TORONTO, ONTARIO, 1970

An individual can hear sounds as music (enjoy living) whether or not he is at a concert.

—John Cage, *M: Writings '67–'72*

A few years later, Bob and I made a visit to John at his house in Toronto. By that time John's bell collection was prominently organized in one room. Each bell was individually suspended to allow for its sustained ringing. Clusters of small bells were also suspended on strings of four- to six-foot lengths in such a way that when activated by a push of the hand, they could swing freely and strike each other for as long as several minutes. In fact, even after it seemed to us that the ringing had stopped, two bells might later come together, producing a single delicate sound completely unexpected and out of context, maybe punctuating a conversation and prompting all of us to laugh.

John was completely enamored of the bell sounds, and he would move around the room playing them while encouraging his visitors to join in. The experience of listening deeply to the sounds of these bells had a lasting impact on our musical sensibilities, as did the experience of using our intuition in creating our own musical responses. Two new conceptual approaches in listening to the bells emerged: there were no wrong sounds, and there were frequent, but seemingly random, occurrences of what might be called *consonance*—a coming together of pitches or timing or resonance. Furthermore, the bell sounds were fully capable of involving us and sustaining our attention.

Inspired by John's bells, the rest of us began collecting our own instruments—mostly bells and gongs. At that time such instruments were to be found only after a considerable expenditure of time and effort in hunting for them in antique shops. We searched in antiques shops rather than music stores because most of the kinds of instruments that interested us had originally been transported to North America by international travelers. Over time, as estates were liquidated, the instruments had found their way into antique shops, where they could usually be purchased at relatively low cost.

ROCHESTER, NEW YORK, 1971

If the goal [of music] is to appeal to a certain number of people, there's a danger of not being true to yourself. Each person has to present the music that's true to one's self, or there's a phony aspect to it.

—Russell Hartenberger of NEXUS, interviewed by the author in 1998

In May 1971, at the prodding of composer Warren Benson, the first NEXUS concert was presented in Kilbourn Hall at the Eastman School of Music in Rochester, New York. The stage was filled with our collection of instruments, which had grown to include not only bells and gongs from all over the world but also our own homemade percussion instruments, and even a few flutes and string instruments. The music in our two-hour concert was entirely improvised. There were two reasons for this. First, we had collected many instruments, the sounds of which we were still interested in exploring, even in front of an audience. Second, every instrument in the collection was one-of-a-kind in its sound (pitch, timbre, resonance) and there were no composed pieces of music available for our specific collection of sounds.

The improvisations that occurred in this concert had no preconceived rules or plans. Every one of the performers was free to play anything at any time. Rather than being a license to "go crazy," the absence of a plan was taken as a responsibility to listen carefully and to make interesting music, just as had been done with the bells at John's house.

The concert was a great success for NEXUS, though we may not have fully conceptualized yet what we were doing. We were not following the established concert canon, in which the performer, speaking by proxy for the composer, says to an audience, "Here are our musical ideas, which are now being presented to you after much hard work and preparation."

NEXUS was instead saying to this audience, "We don't know yet what musical ideas we will be presenting to you; we're going to search for them now, right before your very eyes and ears." Left unsaid was that there was no guarantee. Even if we were successful in finding and developing our sounds and ideas, an audience might not find them interesting. Our assumption was simply that an audience could be as involved in the searching process as we were. Little thought was given to any risk we might be taking.

This approach to playing our music was vindicated when the music reviewer for the Rochester *Democrat & Chronicle* wrote, "How [the performers] germinate, grow, then gyrate a musical idea—from a few fundamental sounds into a sonorous symphony orchestrated with spontaneous, original composition on-the-spot—is utterly fascinating."

For the first few years of its existence, every NEXUS concert was completely improvised. Over time it became apparent to us that the state of mind that existed in our improvisations could be beneficially transferred to making music in other contexts, notably in symphonic music and composed chamber music. Such a state of mind has the following characteristics:

- A deeper knowledge of the instruments and their sound-making possibilities
- A deeper level of listening—to one's self and to other ensemble members—focusing on an acute awareness of the sounds being made
- A more developed intuitive sense in making appropriate musical responses
- An increased ability to embrace the sounds produced by others
- An increased confidence in musical expression and risk taking

The value of our experiences in improvisation gradually became evident to us as time went by. The musical mind-set that has just been described was easily capable of being transferred from improvised music to other areas of music performance. In my own situation it was transferred to the performance of orchestral music.

YORK UNIVERSITY, TORONTO, ONTARIO, 1973

I have experienced the feeling of becoming the actual sound I have been playing … the feeling of literally losing your identity. This is why I don't think the issue is whether the music is improvised or written, Bach or Takemitsu. If that kind of thing can happen, that is music.

—Robin Engelman of NEXUS, quoted in the *Toronto Star,* March 7, 1982

The fascination with improvised music soon motivated NEXUS to present improvisation workshops, starting with a week of improv sessions and concerts at York University in July 1973. From that event onward, such workshops were regularly presented by NEXUS, mostly at universities and music schools. However, a formal pedagogy for improvisation was never developed.

The workshops were structured very loosely. Normally, NEXUS would first perform a short improvisation or two on the group's collection of world percussion instruments and then selected participants would improvise on the same instruments, sometimes joining NEXUS and sometimes without NEXUS playing along. Standard orchestral percussion instruments like marimbas, vibraphones, timpani, and the like were rarely used in the NEXUS improvisation workshops. The general approach to playing was simply for the participants, who were usually percussionists—rarely other instrumentalists, although occasionally there might be a wind or string player—to play whatever they wanted to play.

Around the time of the York University workshops, NEXUS also crossed paths with Paul Winter and his consort of wonderful musicians, all of whom were great improvisers. Soon Bob Becker, Russell Hartenberger, and I were performing with the Winter Consort and interacting with wind, string, and keyboard instruments in the creation of spontaneous music, some of it tonal and some completely free form. The transition from our NEXUS improvisation mind-set—derived mainly from experiences with percussion instruments—to creating improvised music with musicians on other kinds of instruments was effortlessly accomplished.

ATSUGI, JAPAN, 1998

Don't make images: Create meaningful rituals.
Don't occupy space: Identify with it.

—Frederic Rzewski, quoted in John Cage, *Notations*

In 1998 I was invited to Japan as a visiting artist in residence at the Showa College of Music and the Arts near Atsugi, south of Tokyo. Upon my arrival it occurred to me that by improvising with individual students in their initial session in my studio, it might be possible to create an immediate personal bond based on the shared creation of music. I decided to obtain an audiocassette recorder and sound system for my studio so that I could record improvisations for playback and analysis with each student. Instead of using bells, gongs, and other non-Western percussion instruments, I decided that we would improvise on instruments that were not only available, but also of particular interest to many of the students. Each student would play on a marimba and I would use a vibraphone.

For me, this was a big step into new territory. Most of my experience in improvising with NEXUS had been in playing on percussion instruments—drums, bells, gongs, wood blocks, cymbals, rattles, and "found" or homemade instruments—most of which had a clear pitch, but when assembled collectively did not result in any formal scale system or implied tonality. In the NEXUS scenario, there had been virtually no concern about harmonic or melodic clashes. Every sound could be quite acceptable in combination with any other sound.

Almost by definition, the NEXUS improvisations had a certain abstract nature to which percussionists are no stranger. In fact, part of the formal training of a percussionist in music schools is in learning to make subtle distinctions between sounds in order to have just the right sound—say, of a woodblock—for a particular passage in the music. It's only a short step to go from such a conceptual foundation to a full-blown acceptance of sounds alone, regardless of their context, as a basis for the creation of music. This is exactly the kind of step taken by John Cage in his compositions of the 1930s and 1940s. It was the way NEXUS came to approach its music too.

But now in Japan we would improvise on a marimba and vibraphone, which are tuned in the universal chromatic system used for the piano. This opened up the possibility of new risks in playing harmonic or melodic clashes that might sound unacceptably harsh compared to the music we hear around us daily. Our improvised music might be contrary to centuries of accepted harmonic practice in Western music paradigms. Although I was predisposed because of my experiences with NEXUS to accept the idea that any of those twelve pitches on the vibraphone could be combined with any other pitch on the marimba to make music, there was still a specter of doubt about how the music would be received by the Showa students or even by me. But through all of my doubt and the probable doubts of the students too, each improvisation proceeded to be realized.

The responses from the students upon hearing the playback of our improvisations proved to be gratifying in the extreme. Facial expressions were wide-eyed and unbelieving, provoking ear-to-ear smiles. In NEXUS I had become accustomed to listening to recordings of our improvisations, but this was the first time I had observed the power of such listening in others. It brought to mind memories of listening to recordings of improvisations that Bob Becker and I had made in the days just before NEXUS was formed, when we would just gather a few of our instruments and tape record an improvisation. Then we would listen to the playback in amazement that the music was so engaging for us to hear.

Listening to the playback of each improvisation at Showa was followed by a brief questioning and discussion about what we heard. "What did you think about when you were improvising?" "What did you think as you were listening to the playback?" "Did you notice anything in the playback that you didn't hear while you were playing?" "Who was leading in the first part of the piece?" The residency culminated in a final recital in which a group improvisation was included.

Subsequent visits to Showa in the following years reinforced the perception that improvisation—particularly free-form improvisation—has immense potential as a pedagogical tool, for performers, teachers, and students, regardless of the musical genre or style. It has also become evident that the experience of playing free-form improvisations is made more meaningful for players by the additional steps of listening to the playback of the improvisations and participating in a question and discussion process.

By October 2000 more opportunities occurred in Germany, the Netherlands, and at the Banff Centre for the Arts in Banff, Alberta, for me to work with conservatory students and professionals in developing and experimenting with ideas about improvisation.

ROCHESTER, NEW YORK, 2002

In 2002, at the request of the Eastman School of Music, John Wyre and I presented a weeklong summer course titled Improvisation for All. During that event, a simple pedagogy for free-form improvisation began to crystallize. The climax of that week was a wonderful improvised concert performed by the sixteen participants, which not only confirmed the value of this simple method of approaching improvisation but also inspired the writing of this book.

2

QUESTIONS, QUESTIONS, QUESTIONS!

Creative people know that the quality of their products is entirely dependent on the quality of the questions they ask. Skillful inquiry includes seeking and trying good, and then better questions, as well as ongoing answering. This is the most direct route to understanding.

—Eric Booth, *The Everyday Work of Art*

WHAT IS THIS BOOK ABOUT?

Creative music making (CMM) is largely the result of an ongoing search for answers to basic questions:

- Why do I want to make music?
- What is it about making music that I really like?
- What can I do to further develop the things I really like?

These are questions for musicians who want to make their commitment to music even more gratifying than it already is. It is the intent of creative music making to provide a practical method of attaining that end.

It is not uncommon in the education of musicians at all levels for there to be so much emphasis on acquiring technical skills that concerns about individual expression and personal fulfillment—the very things that often inspire the pursuit of music in the first place—are overlooked or neglected.

The kinds of questions that are currently the focus of much attention in learning to perform music are objective:

- How well do I play my instrument?
- At what level are my technical skills?
- What competitive ranking or grading will I receive in relation to others?

At the periphery, if considered at all, are questions such as:

- What do I have to say (musically)?
- What unique qualities of my musical ideas need to be developed?
- How can my involvement with music be channeled into a lifelong enrichment in the quality of my life?

For those in pursuit of a professional career in music, especially those in collegiate music programs, the search for institutional support in the development of individual creativity through spontaneous and uninhibited music making can be daunting. If there is an imbalance, it is generally because there are few, if any, courses to be found in the curriculum that have as a sole objective the nurturing of each musician's individuality through the reinforcement of spontaneity, introspection, and personal expression in music making. It is revealing that there seems to be very little institutional awareness of such a void.

WHERE ARE WE AND HOW DID WE GET HERE?

The trend ... from complexity to simplicity, has commonly been hailed as a healthy return to musical grass roots, as a kind of intellectual and esthetic sobering up.

—Henry Pleasants, *The Agony of Modern Music*

The idea that music can and should have a profoundly positive impact on our lives has existed throughout history. The conventional wisdom is that the ability to understand and appreciate music is an essential component in a well-rounded, liberal education. The study of music is considered to be an important tool in the development of abilities as diverse as critical thinking, intuition, motor skills, social skills, and creative problem solving.

It is also widely perceived that the study of music can foster a stable balance between the rational, logical aspects of our consciousness and the nonrational, intuitive, and spiritual aspects. It is the formation and preservation of such a balance that is the subject of concern here. Over the course of the last few decades in North American music education, there have been significant shifts of emphasis that have affected this balance.

In the 1930s and 1940s, with widespread access to radio and electric phonograph recordings, which are both aural, nonvisual media, more people could listen to music than ever before. Listening to music was considered by educators to be a skill that could be developed by everyone through regular practice. For performers, it became possible to have greater access to the full range of the world's repertoire of musical ideas.

In the 1950s and 1960s, as visual mass media gained in prominence, the balance of active listening and active performance skills for music students, from grade school to conservatory, gradually shifted in favor of performance. In music schools and conservatories, students' time was increasingly devoted to acquiring and perfecting performance techniques. The development of listening skills was increasingly limited to music theory classes.

By the end of the 1990s, with a profound revolution in technology and communications underway in computers, the Internet, digital imaging, digital sound recording, and so on, the education of performers focused increasingly on objective concerns—students' technical competence in performance, in competitions, and in juries—within a system of measurable standards set by organizations of professional musicians and educators.

Subjective musical concerns—individual musical expression, spontaneity, and soulfulness—have been diminishing concerns in this environment, perhaps overwhelmed by such a widespread fascination with newly accessible technical marvels.

It cannot be denied that the emphasis on playing techniques, especially in the study of instrumental music, has raised the general level of performance to ever-greater speeds and complexities, which is a good thing. However, the question arises, is that all there is?

IS PERFECTION THE GOAL OF MUSIC MAKING?

> *Perfection increases as inspiration decreases ... a useful corollary is that perfection is not a necessary characteristic of the greatest art.*
>
> —Jacques Barzun, *From Dawn to Decadence*

In a world where most music is heard electronically in highly edited, "perfect" performances, such high technical standards place a considerable burden on live performers. The technical achievements made possible by the electronic media are rarely, if ever, attained in live performance, because people are simply not perfect. The inherent risk for performers in such an environment is that the pressure to be technically perfect will have a dehumanizing effect if not balanced by humanizing concerns and activities.

This problem suggests other fundamental questions:

- Is music to be valued for its positive effect on participants—listeners and performers—or is it to be judged solely on the competence of the performers in rendering a technically perfect performance?
- Can a definition of "music" be preserved that takes into account some measure of an imperfect, yet creative and engaging musicianship?
- Is there any room left for spontaneity, unpredictability, and surprise in music?
- Can a sense of spirituality in music performance be preserved?

Yes, yes, yes, and yes, of course! The method recommended in this book is intended to address such issues. It is intended to provide some degree of balance for musicians in the psyche of music making. Its objective is to provide a practical means for musicians—whether professional, amateur, or student—to develop their own individual voices and to widen the spectrum of their musicianship. If at the same time it enables musicians to take greater advantage of the full range of music's powers—to foster a love of life, to enhance an understanding of one's self and of one's connectedness to others, and to reinforce one's sense of fulfillment through the exercise of personal creativity—so much the better.

3

A PEDAGOGY FOR MUSICIANSHIP

... but, where is the music?

—Pablo Casals, as overheard by NEXUS members
at the Marlboro Music Festival, 1968

Why is it so difficult to find courses of study in music that are devoted primarily to the basic concerns of musicianship—good listening, creative expression, and the exercise of musical intuition? Among other factors, the high priority given to technical concerns, which is occurring all over the world and especially in North America, is a result of the following scenarios that can be found at all learning levels in music:

- There is a high value on competition—winners and losers. This has been exacerbated by the ever-growing number of students seeking careers in music. In North American culture, a competitive environment is largely influenced by the overwhelming presence of sports paradigms. It is always difficult at best to judge performers based on subjective criteria, because the inescapable personal prejudices of the judges can be as influential in the judgment as the competitors' actual performances. Technical criteria—for example, the execution of the written notes in performance—are simply easier to use as a basis for making judgments about a performer.
- There is a high value placed on grading—relative standing in comparison to others—also possibly driven by analogy to sports: this is exemplified by the grading standards to be found in many states for precollege music students and by formal juries at the college level. Again, subjective concerns can only produce relatively weak comparisons—apples to oranges. Technical evaluation is more likely to be objective, and as such, is also more likely to take precedence as a factor in determining grades. Further, it is not uncommon for the grade itself to become the goal rather than the underlying accomplishment that the grade is supposed to represent. This may result from institutional or social pressures on the teacher as well as on students.
- There is a complete absence of an accepted pedagogy in support of the nontechnical elements of musicianship.

This last point is one that will be addressed in this book. The absence of a pedagogy for musicianship is understandable. The subjective nature of musicianship is in itself an impediment to any attempt to objectify it through the structure of a system of pedagogy. But without such a system, institutions of music have not taken subjective musicianship seriously enough to include its study as a formal part of the regular curriculum, preferring instead to leave subjective matters to the discretion of the private studio, where the issue may or may not be addressed.

Except for the study of music that specifically incorporates "idiomatic" improvisation (jazz, figured bass, North Indian, African, Latin American, and so on), improvisation —particularly free-form improvisation—as a vehicle for addressing the subjective side of music performance has been all but excluded from formalized institutional support largely for the same reason: the absence of a widely accepted pedagogy.

For institutions largely concerned with the objectification of music making, free improvisation is a subjective anathema. It is ephemeral and elusive to analysis. For some in academe, improvisation—especially free-form improvisation—may even be dismissed altogether as a form of disordered self-indulgence that fails to rise to institutional standards for music making.

A pedagogy of improvisation is anathema even to many musicians who regularly improvise. It is widely thought that attempts to create a structure for improvisation are doomed to fail because external structures imposed on spontaneous performance will necessarily transform its nature by restricting it in some way.

For musicians who do not normally improvise, such as symphony orchestra players, there can be a considerable amount of discomfort in taking on the responsibility for actually creating the musical material. This is because in certain kinds of classical music—for example, "contemporary" or "modernist" art music—there may be no generally established improvisatory style. If a composer indicates that an improvisation is to occur, it may not be possible for the performer to know exactly what the bounds of such an improvisation should be unless the composer has indicated such boundaries in an accompanying text. The unpredictability of the consequences of either doing too much or of being too cautious is what can produce discomfort—a state of mind that, to say the least, is not ideal for music making.

Unfortunately, the word *improvisation* has become so weighted down by this kind of conceptual baggage that it is tempting to avoid the term altogether in order to avoid negative prejudices and to focus more on the positive effects of its application within a system of pedagogy.

For that reason, the simple pedagogy presented in this book is not called free-form improvisation. It is called creative music making (CMM)—a formal structure that is not in any way restrictive of musical content. Rather, it is a structure for the regular exercise of musicianship, regardless of the specific musical content. The four-step CMM process is simple enough to be summarized on a single page and complex enough to require protracted study.

4

MUSIC MAKING ... FOR WHOM?

> *In music, the idea of specially gifted individuals is virtually axiomatic in Western cultures reflecting their emphasis on the individual. ... But, as is now well known—at least among anthropologists and ethnomusicologists—the innate talent view of excellence is hardly a cultural universal ... in a number of traditional cultures virtually everyone has a palpable degree of professional music status.*
>
> —Jeff Pressing, "Psychological Constraints on Improvisational Expertise and Communication"

Creative music making is a method of finding and remaining in contact with the inner musician—the spirit that was touched by the expressive power of music in the first place. CMM can be performed on any musical instrument or combination of instruments, including the voice. Strings, woodwinds, brass, percussion, keyboards, electronic instruments, folk instruments, found or handmade instruments in any combination—all can be utilized.

The four steps of CMM can be practiced at any level of involvement in music. Although CMM may be of particular value to professional musicians and career-track music students, there are benefits to be found in CMM for almost anyone who wants to learn more about music or wants to play a musical instrument. It can also open the way to a deeper understanding between music teachers and their students.

CMM can have a profound impact on the following categories of potential practitioners.

CAREER-TRACK MUSIC STUDENTS

First and foremost, it is career-track music students who can benefit from CMM sessions on a regular basis, not only as a way to deepen the experience of performing but also as a means to engage in a regular musical dialog with teachers and fellow students.

In many music schools, much attention is given to the development of virtuosity as a soloist, even though most musicians will ultimately become performers in an ensemble. CMM is a way of creating confidence for the performer in interacting with other ensemble members. It provides an environment in which it's possible to experiment with musical ideas and discover their effects on other players and listeners. In an atmosphere of playful trial and error, CMM enables the student to obtain valuable feedback from others. It enables aspects of each performer's own musical sensibilities to come into greater focus, so they can be applied in the performance of any kind of music.

CMM also opens up the possibility of making improvisation an integral part of the student's repertoire. With experience in the CMM process, students can present their own free-form improvisational pieces in the context of almost any concert program.

The enclosed CD starts with two wonderful examples of improvisations performed at a public recital titled Improvisation for All that was presented in Kilbourn Hall at the Eastman School of Music in Rochester, New York, on July 19, 2002. None of the student performers on these two tracks had ever performed a free-form improvisation prior to their participation in the preceding week of daily creative music making sessions.

Play CD Track #1—Improvisation Duet (Time: 5:21)

This is a violin and piano duet played by Ko Taniguchi and Seung Hae Jung, who were college-level students at the Eastman School's summer session in 2002. Both players clearly displayed their formal training as career-track music students with a strong background in classical music. The resulting improvisation has a classical sound, reflecting the vocabulary of the performers.

As frequently happens with free-form improvisations played by experienced musicians, many listeners, upon hearing this duet, expressed their complete amazement that it was not a formally composed piece. And, both players had never performed in public together before this improvisation.

Play CD Track #2—Improvisation Quartet (Time: 6:55)

The music in this free-form improvisation is performed by Eastman summer session 2002 participants Ya-ting Lee and Huei-hsien Wang on piano, and Jin Kim Soo on violin. All three were college-level conservatory students. Also playing in the ensemble on percussion was Edith Mann, an adult nonprofessional participant in the week of CMM workshops. The music is a great example of what can be done when players at varying levels of experience perform a free-form improvisation together in concert.

PROFESSIONAL MUSICIANS AND MUSIC TEACHERS

Professional musicians and music teachers can also benefit greatly from experience in CMM sessions. Free-form improvisation can be a means of expanding an already established musical vocabulary. The reinforcement of confidence in one's intuition can be a significant enhancement to an already active musicianship. The spontaneous sharing in the making of music can be a powerful means of engendering a deeper communication with colleagues, audiences, and students.

Instrumental and vocal teachers can use CMM in the studio as a technique to share musicianship with the student and to learn more about the student's musical sensibilities.

GENERAL MUSIC STUDENTS

Music students, at any level, whether on a career path or not, can also benefit from the regular practice of CMM. Among the desired outcomes of such exposure would be the ability to perform music with imagination and the confidence to explore one's own unique musical ideas using ever-developing technical skills. Another benefit would be the ability to have an appreciation for the musical ideas of others and to make listening to music a lifelong source of enrichment.

AMATEUR MUSICIANS

Amateur musicians can certainly gain important music-making perspectives from regular CMM sessions. CMM can enable the amateur musician, even without a high degree of technical skill, to achieve a level of musical expression in performance comparable to that of professional musicians. It can also develop the listening skills that can make it possible to hear unfamiliar music and to understand it in a meaningful way.

NOVICE MUSICIANS

Technology already makes it possible for virtually everyone—the mass market—to engage in CMM-like activities on a regular basis, even though the practitioners may not be aware that they are doing so. The grade school child or office worker who plays on any inexpensive electronic keyboard can create tunes and play them back on the inboard sequencer for immediate evaluation. These keyboard instruments enable the player to have an acceptable sound right away, which shortens the time that would otherwise be needed to overcome sound production problems before actually getting to play satisfying music. However, one important aspect of music making is usually missing for anyone using this technology: it is interaction with other musicians. CMM can fill in this gap.

Novice musicians of any age can (and should) also participate in CMM. If they choose to play on standard band or orchestra instruments, it will be best if they at least have the basic skills to produce an acceptable tone on their chosen instruments. In the absence of such skills, there are other instruments available that can be played using simple techniques.

For example, on a piano, the novice and even the nonpianist professional can achieve an acceptable sound almost immediately using a simple technique, playing with only the index fingers, which will enable the player to explore melody, rhythm, and dynamics readily.

Resonant keyboard percussion instruments like marimbas or vibraphones are also recommended because they can be played melodically and they only require a minimal amount of technique to produce a fairly good sound. The basic playing technique on any keyboard percussion instrument is simple:

- Relax; hold a pair of medium yarn-wound mallets—one in each hand—comfortably, with no tension in the fingers or hand
- Strike the wooden (marimba) or metal (vibraphone) bars just off center, using a relaxed "lifting" or "rebounding" stroke

Fig. 4.1 One kind of marimba, the homemade amadinda, is ideal for novice and professional musicians alike in CMM improvisations.

Xylophones and glockenspiels, though also categorized as keyboard percussion instruments, have a higher register and a harder, less resonant sound, so they are not recommended for novices. The operative assumption here is that resonant sounds are generally more engaging.

One kind of marimba, the homemade amadinda, is ideal for novice and professional musicians alike in CMM. This instrument is a large, wood-board marimba based on a type of African xylophone found in Uganda that is called an amadinda. The homemade version of this instrument is easy to build using materials that are widely available. Instructions for building a homemade amadinda are given in the Building a Homemade Amadinda section.

One advantage of the homemade amadinda is that its musical scale is likely to be irregular and unfamiliar sounding. Consequently, the urge to try to play familiar tunes—which happens easily on the piano or keyboard percussion instruments—is reduced and the freedom to explore original musical ideas is encouraged. Another advantage is that more than one player can play on an amadinda at a time with one or two players on each side of the keys.

Play CD Track #3—Amadinda Improvisation (Time: 7:05)

Included on the enclosed compact disc is an excellent improvisation on a homemade amadinda, performed by high school students Sebastian Henshaw, Edith Resnick, and Jordan Schifino, with Bevin Coggeshall, a first-year college student in percussion. Although these students knew each other prior to the CMM workshops at the Eastman summer session 2002, this was their first improvisation in front of a live audience.

NONMUSICIANS

Even nonmusicians, young or old, can participate in CMM and experience some of the profound joys of making music—finding and connecting with oneself and with others. For example, adults in a continuing education music appreciation class might "play" music on instruments found in the music room or possibly on instruments—like the homemade amadinda—constructed by the participants themselves. Even without any acquired technical skills, the simple ability to play a stroke on a percussion instrument or to play on a piano with the index-finger technique may be all that is required to begin to explore music's powers through the CMM process.

FAMILIES

The sharing of music through singing and playing in the home with family members is a tradition that has all but disappeared. However, for any family that might be so inclined, CMM can provide a useful structure to enhance the experience of regular music making together in the home.

5

THE BASICS OF CMM ... JUST DO IT!

[Art] is not a product; it's a living process ... [Ananda K.] Coomeraswamy said, the artist is not a special kind of person, but every person is a special kind of artist.

—Satish Kumar, quoted in Suzi Gablik, *Conversations Before the End of Time*

WHAT IS CREATIVE MUSIC MAKING?

Creative music making is a four-step process in which musicians can expand their musical expression through the creation of spontaneous, free-form music without the constraints of thinking about technique or following printed music. As previously indicated, any musician or music teacher of any age, student or professional, performing on any instrument, will find CMM to be a practical method of cultivating his or her musical voice. At least two performers, not necessarily at the same level of experience, are needed to participate in a CMM session.

WHAT ARE THE FOUR STEPS OF CREATIVE MUSIC MAKING?

Step 1: *Playing*—participants create spontaneous, free-form music in small groups of two to six performers; no previous experience in improvisation of any kind is required.

Step 2: *Recording*—participants have their music recorded for playback, listening, and analysis.

Step 3: *Listening*—participants focus on improving listening skills—noticing as much as possible about whatever is heard—as a means of expanding musical vocabulary.

Step 4: *Questioning*—participants, in a group round-table format, raise and respond to appropriate self-directed questions in order to increase understanding and gain confidence.

WHAT ARE THE RULES FOR PLAYING FREE-FORM IMPROVISATIONAL MUSIC?

There are only two things for participants to know before starting to play in a free-form improvisation; for the purposes of creative music making they will be called "rules." But these rules are intended to be open-ended and liberating rather than limiting.

Rule 1: Performers may play (or not play) anything they wish on any available instrument of their own choosing—there are no mistakes.
Rule 2: Performers should listen as deeply as possible to themselves and to the other performers, but it's important that it be perfectly clear to all participants—players and listeners alike—that there is no penalty for breaking this rule.

IS ANYTHING ELSE NEEDED FOR CREATIVE MUSIC MAKING?

Actually, having read the preceding steps, you now have all of the basic information you need to begin CMM. Also needed will be fellow participants with whom to make music, and musical instruments that you like to play. The help of a facilitator in addressing the concerns of new participants will likely be needed as well, particularly in the first few sessions.

Yet before these other needs are examined in greater detail, it will be worthwhile to consider two related underlying aspects of music making that CMM seeks to address—creativity and intuition.

CREATIVITY

I define creativity as the ability to consistently produce different and valuable results. ... By this definition, creativity is an ability that can be improved; it does not belong to any one particular group of individuals, born with a certain, special set of traits.

—Lynne C. Levesque, *Breakthrough Creativity*

Regardless of the genre of music, or of the techniques used to realize it, every act of music making is a creative act. Active listening is also an act of creation. The performing musician starts with silence and makes meaningful sounds. The listener starts with silence and makes sounds meaningful.

Contrary to the common notion that creativity is a mystical quality possessed only by gifted individuals, creativity can be viewed as nothing more than a commitment by anyone to make something. For the purposes herein, it is assumed that the ability to make a sound is an ability that everyone possesses. It is also assumed that everyone is capable of attaching meaning to the sounds made—to make sense out of sounds that may at first seem to be disordered.

The main ingredient necessary for creativity is the determination to make something —in this case, sounds—even if there is uncertainty about how meaningful those sounds will be. An inability to accept this uncertainty can be a significant obstacle to action, blocking the act of making. A fear of making music that might be judged as less profound than a symphony by Ludwig van Beethoven can be very inhibiting.

Yet if a musician really wants to "play" music, then an attitude of play—not unlike a child's—will be very enabling. To play is simply to act for the pleasure of doing. Whatever happens is what happens. It is in the responses to the things that happen, whether expected or unexpected, that it's possible to learn and grow.

The playing of music and the attaching of meaning to it may be related but they are still two separate actions. First, make the sounds—the music; in other words, create. Then, seek meaning. Ask questions and make connections among yourself, the music, the other players, and any other aspects of the experience that come to mind.

INTUITION

Despite the widely-held assumption that … deliberative, logical, analytical thinking … represents the most powerful thinking tool we possess—which makes it the one we call upon, or revert to, in the face of urgent demands for solutions—the truth is our ideas, and often our best, most ingenious ideas, do not arrive as the result of faultless chains of reasoning. They "occur to us." They "pop into our heads." They come out of the blue. When we are relaxed we operate very largely by intuition.

—Guy Claxton, *Hare Brain Tortoise Mind: How Intelligence Increases When You Think Less*

These pages outline a simple system of study for the development of creativity and self-expression through regular opportunities to "play" music and to educate the player's intuition. A well-exercised sense of intuition opens the door to greater spontaneity and freedom in music making. It also enables the emergence and cultivation of a personal voice—a truthfulness to oneself—and an ever-increasing satisfaction in music making, at any level of technical ability.

Intuition is a nonrational (but not *irrational*) means of selecting from among the numerous possibilities that every musician necessarily faces in performance—what note to play, how loudly, what tone quality, what shape or contour of sound in the context of the ensemble, on and on. *Nonrational* does not mean that there is no thought involved. Thought may or may not be just one of the means used in determining which of the many performance options is actually selected. Other possible means of selecting might be through imagination, memory, or emotional state, for example.

In performance, so many selections are possible that there is not enough time between the notes to think through all of them. Additionally, there is some risk involved: What if I make a poor selection?

It can be very tempting to deal with this concern by trying to eliminate the need to make such selections in performance altogether. This, in a way, is what technically "perfect" recordings of music do. Except for the actual playing in the recording session, recorded music sterilizes against the possibility of spontaneous selection, every single time. It virtually eliminates the possibility of any unpredictable occurrences in the performance.

One common method used by performers to eliminate the real-time selection of performance options is to think through and "fix" as much as possible in practicing, so that in performance the music will be so automatic that no selections need to be made. In this way the selections are all made in advance and not in the real time of the performance.

Many performers view this method as the preferred way to approach music making. For professional musicians, it is a practical method of dealing with the normal pressure to be as flawless as possible in their performance, and as such it can be useful and effective.

Yet, another risk appears. This method can also produce a "studied" performance, lacking in spontaneity. Why? Because spontaneity is simply the selecting of musical elements (by whatever means) from among the many possible ones in real time—selections influenced not only by advance planning, but also by the immediate performance as it happens. It is spontaneity that gives each performance its unique identity and can make it possible to reach deeper levels of communication in the experience of music making—both for the performer and the listener. Its presence is what gives real "life" to a performance.

Instead of trying to eliminate real-time musical selection in performance, why not embrace it by regularly including it in practice sessions? Selecting in performance by means of intuition needs to be regularly practiced, too, so that it can be done naturally and with confidence.

6
IMPROVISATION

> *Music can never be accidental, however improvised ... because improvisation is not the expression of accident but rather of the accumulated yearnings, dreams and wisdom of our very soul.*
>
> —Yehudi Menuhin, *Theme and Variations*

Creative music making is, among other things, a good method for regularly exercising musical intuition. In many ways it is similar to free-form improvisation, but CMM is much more. In order to compare them, it will be helpful to first consider what is meant by improvisation in music.

According to *Webster's New School and Office Dictionary*, the word *improvise* means "to compose extemporaneously; to bring about without previous preparation; to devise on the spur of the moment; or to do a thing in an offhand way."

Each of these definitions carries with it certain implications. "To compose extemporaneously" or "to devise on the spur of the moment" means to make something in the present moment. However, the present moment does not exist in a vacuum. It did not come into existence from nothing. It is derived from all of the moments that have preceded it.

"To bring about without preparation" means to make something happen without planning for it in advance. However, what about all of the moments that preceded the improvisation—whether or not there was any conscious planning related directly to the music created? Didn't those moments have some effect on the result?

"To do a thing in an offhand way" may imply that the thing that is improvised has been treated frivolously, with less care or concern than would have been given with some planning. However, what if the improvised music came from a desire to intensify the act of making, rather than from some frivolous whim? Is it likely that such a desire would be treated with a lack of care and concern? Not if the act of making or creating is valued.

It is precisely when the creative act is valued that improvisation has a role to play. The creative act is valued because it is a powerful way of embracing the activities of life and of giving them meaning.

Improvisation need not be limited to the arts. It can reasonably be argued that most activities, musical or otherwise, involve some degree of improvisation—deciding what to say in a conversation, choosing clothes to wear, or even selecting a wine at dinner. Approaching everyday activities with an attitude of improvisation is one way of deepening involvement in life and discovering more about oneself.

Music is not the only art form in which the ability to improvise can be a valuable skill. In the study of theater arts there are formalized courses in improvisation, because the ability to improvise with confidence in acting or directing is highly valued. Actors performing in auditions are routinely asked to demonstrate their abilities in improvising short scenes. Improvisation in dance is also valued for its capacity to produce a "centered" consciousness in which dancers can find and remain in touch with their true spirit and physical abilities.

Improvisation also works as a means of intensifying the experience of making and listening to music, and it is to be found almost everywhere music is being played in the world. And yet improvisation is typically not included as a formal part of training for most musicians in music schools, except maybe for those who study jazz or perhaps baroque ornamentation and figured bass.

In the music of North America, improvisation exists in differing forms, depending on the type and style of music. In classifying the forms of improvisation, Derek Bailey has identified two primary types in his excellent book *Improvisation: Its Nature and Practice in Music.* The first type is *idiomatic improvisation*, which is to be found in most musical idioms—Western and non-Western—like baroque, jazz, African, Indian, Flamenco, Balinese, Puerto Rican, and so on.

In idiomatic music there are well-defined stylistic rules that apply to improvisation. The rules in any particular idiom may be written down or unwritten, but the requirements and limits of improvisation are generally accepted and understood by virtually all performers and listeners in that idiom. Even so, it is quite possible for the rules to change over time as styles evolve under the influence of successive generations of performers, but any improvisation that is not within the generally prescribed boundaries will be recognized immediately as such by listeners familiar with the idiom.

However, it is not the intent here to present an analysis of all of the prescribed methods of improvisation that are practiced in the numerous genres of idiomatic music to be heard in North America, let alone the rest of the world.

For the purpose of better understanding the way in which improvisation is practiced in creative music making sessions, it will be sufficient to make a few key points about the most widely known genres of idiomatic improvisation in North American music.

IDIOMATIC IMPROVISATION IN JAZZ

> *A musician just has to learn for himself, just by playing and listening. ... There ain't no one can write down the feeling you have to have. That's from inside yourself. The music has to let you be ... you got to stay free inside it.*
>
> —Sidney Bechet, quoted in Nat Shapiro, *An Encyclopedia of Quotations about Music*

The genre of music in North America that is most often associated with improvisation is jazz. In jazz improvisation, musical elements like melodic lines, harmonies, rhythms, dynamics, and the interplay with other musicians are all open to the performer's real-time selection. It is this fact—that the musical details are being selected in real time, as the music is being created—that gives the music much of its power. The selections that are made are what also reveal each performer's individuality.

However, to say that jazz improvisation happens "on the spur of the moment" or in an "offhand" manner could be very misleading, because such descriptions do not adequately take into account the accumulated experience that is so essential. In jazz improvisation the selection of musical material is profoundly influenced by what has gone before—not only in the immediate performance, but in numerous past performances, by a host of players.

The real-time selection of musical elements is generally expected to be consistent with the jazz composition—the tune, chord progressions, phrase length, and so forth. Also, in jazz improvisation a working knowledge of music theory, standard performance techniques, and musical styles is required of each performer. It normally takes years of study and experience to reach the point at which a jazz improvisation can occur that is satisfying to other players and listeners.

It is worth mentioning that some of the musical elements in jazz—the instrumentation, the form or structure of the composition, or the style—can be rigidly fixed by standard practices in the particular genre of music (Dixieland, be-bop, rock fusion, Brazilian, etc.). Improvisation may not always apply to these stylistic musical elements.

IDIOMATIC IMPROVISATION IN NON-WESTERN MUSIC

In every part of the world we can observe activities that would fail to achieve their ends if performers were unable to improvise effectively in the presence of other participants.

—Stephen Blum, "Recognizing Improvisation"

Improvisation is an integral element in much of the music of the non-Western cultures that have emigrated to North America. Starting as long ago as the seventeenth century with the immigration of slaves from West Africa, music in North America has been subject to non-European influences, including methods of improvising that are particular to each musical idiom.

Today, with the ease of international travel; with the ready availability of the world's musical instruments through global trade; and with instant access to the music of the world through radio, television, compact discs, and the Internet, the rules for improvisation in all idiomatic musical styles are increasingly subject to outside influences. This phenomenon is not necessarily new, but what is new is that the outside influences on idiomatic forms are happening with much greater speed and frequency. Such influences may generate considerable tension between performers who adhere to traditional rules of improvisation and those who incorporate nontraditional ideas borrowed from other genres of music.

Nevertheless, within each idiom stylistic rules remain for improvisation—limits that may be unwritten but are clearly recognized by those who play and listen in the idiom. As in jazz, whether improvising a solo on the santur (a Persian dulcimer), on the sarangi (an Indian strummed string instrument), or vocally (as in modern Latin dance music), the parameters of each idiom can require years of listening and study to understand, and any improvisation must generally be in conformity with the stylistic expectations of the other performers and listeners in that idiom.

IDIOMATIC IMPROVISATION IN WESTERN CLASSICAL MUSIC

> *I would sit down and begin to improvise, whether my spirits were sad or happy, serious or playful. Once I had captured an idea, I strove with all my might to develop and sustain it.*
>
> —Franz Joseph Haydn, quoted in Ian Crofton and Donald Fraser, *A Dictionary of Musical Quotations*

In a live televised broadcast in 2004, Yo-Yo Ma's performance of Franz Joseph Haydn's Concerto in D Major featured improvised cadenzas on the cello, something that was viewed as so unusual that it was given particular attention in the preconcert publicity. Making up music on the spot for an unaccompanied solo incorporating themes from a composed concerto was a practice that existed in Haydn's era and continued into the twentieth century. A soloist's ability to improvise in this manner has been recognized and valued in the past as an indicator of good musicianship. However, by the end of the twentieth century, improvisation in Western classical music performance had virtually disappeared.

It is not widely appreciated today that improvisation has long been a part of Western classical art music, as well as of church music. For an instrumental soloist performing a cadenza in a concerto, every musical element—melody, harmony, rhythm, volume, and interplay with the other performers—still continues to be wide open to the performer's real-time choices, as long as these elements are generally consistent with the composition (melody, motif or theme, harmonic progression, phrase length, etc.).

In plainsong, cantus firmus, or baroque ornamentation, a thorough knowledge of music theory, standard performance practices, and musical style is going to be required of the performer. Such knowledge can only be obtained through years of study and performance experience.

In orchestral classical music, improvisation may take the form of the conductor's spontaneity in interpretation of the phrasing or tempo, or there may be traces of improvisation in individual musicians' phrasing or tone quality, influenced by the spirit of each particular performance as it unfolds. Even so, great care usually must be taken that generally acknowledged boundaries not be crossed, lest an uproar of protest be raised by other musicians, critics, or the general audience.

Of course, in classical music, there are also fixed musical elements that are not subject to improvisation. The instrumentation, structure, and style are generally predetermined by the composer or by accepted practices in the genre of music.

Another type of idiomatic improvisation is sometimes found in contemporary Western art music. It is based on abstract visual or graphic notation. In this kind of improvisation, performers respond to abstract, representational, or indeterminate symbols notated in the score. Here it is entirely possible that instrumentation, form or structure, style, melodies, harmonies, rhythms, dynamics, and interplay with other performers may be left completely to the performer's choices, except as may be specifically instructed by the composer (for example, specific instrumentation), or by the implications of the notational framework, for example, by relative high or low notation symbols implying high or low pitches.

In abstract visual or graphic improvisation, some—or possibly much—knowledge of music theory, standard performance techniques, musical style, and the like may be required of the performer. Again, meaningful improvisation in contemporary art music can require years of study and performance experience to acquire a knowledge of the prescribed practices, as well as of the boundaries or rules of the idiom. It might be completely inappropriate in a piece of contemporary art music for a performer's improvisation to include a parody of a well-known popular song, for example.

Also, as pointed out earlier, in contemporary art music there may be considerable ambiguity about any particular aspect of the composer's intent in asking the performer to improvise. A player might reasonably wonder, is it appropriate to play softly, speedily, using technical effects, quoting themes, using diatonic modes? To add even more risk for the performer, a conductor may not agree with the performer's choices. Consequently, it is not uncommon for classical musicians inexperienced in improvisation to be somewhat hesitant, even downright reluctant to improvise.

FREE-FORM IMPROVISATION

> *In the conception of the art music world, there is a set of parallel contrastive relationships: between composition and improvisation, between crafted and inspired composition. ... But can we tell from the outcome which is which?*
>
> —Bruno Nettl, "An Art Neglected in Scholarship"

The second of the two forms of improvisation identified by Derek Bailey is *nonidiomatic improvisation*. This can also be considered *free-form improvisation*.

In nonidiomatic improvisation there are no limits or parameters for the musical material that can be included. By definition, there are no general rules or constraints of style. Each individual performer is completely free to play music using whatever rules he or she wants to use, with the understanding that such rules apply only to that individual and not to any of the other players.

Free-form improvisation is the most widely open and accessible type of improvisation for musicians. In this kind of music virtually all of the musical elements are subject to the performer's real-time selection. The only structural limitations (in length, motifs, etc.) are those self-imposed by each performer. In its purest form the performers are completely free to play whatever they wish however they wish, with virtually no external restrictions imposed.

The performers are only limited by their own abilities to produce sounds on their instruments, and by their abilities to draw with imagination upon the musical ideas that they have internalized through past listening and performance experiences, regardless of the type of music.

Yet even at the novice skill level there is still plenty of room for creativity. The use of a piano or of keyboard percussion instruments like the marimba or homemade amadinda can enable most novice musicians to play with an acceptably good sound, using simple and easily learned playing techniques.

Because little or no knowledge of music theory, musical styles, or standard performance techniques is required, free-form improvisation is immediately accessible to most musicians, whatever their level of skills and experience.

This is not to say that in free-form improvisation experience doesn't matter. The music created will certainly be affected by the presence or absence of experience. The main effect of experience in a free-form improvisation will usually be the player's greater ability to listen and make appropriate musical responses. It will also be heard in the larger size of the experienced player's vocabulary or in the greater complexity of the musical ideas. But, in the meaningfulness of the music to the listeners, the experienced player's contribution will be equal to that of the inexperienced player.

This kind of music making emphasizes the process of musical communication among the performers, while musical form and structure—critical elements in idiomatic types of improvisation and music composition—diminish in importance. To borrow a phrase from the distinguished composer Toru Takemitsu, in free-form improvisation, the *noun* music is transformed into the *verb* music. Music becomes less of a product and more of a process of doing.

In this kind of music, any conception of "wrong" notes is simply refuted. Fear of playing wrong notes is perhaps the biggest impediment in the psyche of music making, and in free-form improvisation any concept of right and wrong notes can and should be discarded without reservation.

Concern about what is right or wrong is replaced by an effort to be aware of whatever is happening and to search for and find appropriate musical responses—to make good musical choices. It is in this search for responses that free-form improvisation becomes a fertile ground for the exercise of intuition and imagination. The musical responses are derived from each individual's personal vocabulary and experience, without the impediment of fear about playing something wrong.

AESTHETICS OF FREE-FORM MUSIC

> *In improvisation you are living in the moment and making choices based on who you are, where you are, and what you are doing.*
>
> —Greg Atkins, *Improv! A Handbook for the Actor*

It is likely that new participants in creative music making will also be new to the idea that there are no wrong notes. It would not be unusual for a healthy skepticism to be present in their thinking, even if there is a hesitancy to express it. If a new participant has a relatively small musical vocabulary or a narrow scope of taste, maybe consisting entirely of popular or commercial music forms, there might even be hostility to the concept.

For anyone who has not experienced free-form improvisation there may also be some reluctance to become involved in a kind of music where there are no clear boundaries; wouldn't the result just be chaos?

The question presupposes that if there is no perception of order, then the cause is to be found in the object—the music. But if it can be accepted that one person's chaotic music can be another person's beautiful music, then the options for the listener are either to shrug off as meaningless any music that is not understood, or to try to understand what it is in the music that another person might perceive as beautiful or meaningful. In CMM, the second option is preferred.

An inexperienced listener may be very uncomfortable with music that is not programmatic or tonal. In one CMM session, a new participant was even heard to say, "I don't like music without words!" There was nothing inherently bad about this; it was an honest expression that was recognized as such by all of the CMM participants with no need for debate. Rather than try to change the opinions of participants, the goal in CMM sessions is that participants have opportunities to learn and decide for themselves through playing, listening, and questioning.

Just as there can be something of beauty in people who are not movie stars or fashion models—that is to say, in all people, regardless of differing individual characteristics—there can be beauty and meaningfulness in differing genres of music, even in music that is unfamiliar and poorly understood. The key to finding beauty and understanding is to seek it.

At the other end of the spectrum, a new CMM participant might be a very experienced performer or listener, but the thought of "anything goes" may bring to mind negative experiences in playing or hearing contemporary art music, meaning anything from twelve-tone music to the late-twentieth-century "avant-garde." Even within the realm of the cognoscenti, such music may have a reputation, deserved or not, of being harsh and difficult to listen to and understand—in other words, inaccessible.

One of the desired outcomes of CMM is that any inhibiting perceptions and attitudes may be examined, ultimately to be replaced by a new openness to and appreciation for a wider spectrum of music.

CMM is not about making music that is harsh, dissonant, wild, or inaccessible. Just the opposite is true; it is about working together in a process of searching for a musical unity based on careful listening and on inwardly honest playing.

It is certainly possible, or perhaps unavoidable, that in this musical process of searching, a free-form improvisation will lead the players into dissonant or atonal musical territory. If so, that would be just fine, but neither dissonance nor tonal harmony are the final goals. The intent is to learn more about one's self, and to develop a confidence in relating positively to whatever is going on in the music.

The CMM facilitator has an important role in providing as much support as possible in helping new participants to work through any self-imposed impediments. This can be best accomplished simply by maintaining a nonjudgmental environment, allowing space for new participants to absorb their new experiences in improvisation and to reflect upon them later in discussion.

As experience in CMM increases, it ought to become evident to participants that right and wrong notes are not what free-form improvisation is all about. It's about the process—the ways in which the performers and listeners relate to themselves and to others in the context of whatever music is happening.

I KNOW WHAT I LIKE!

I love Beethoven, especially the poems.

—Ringo Starr, quoted in Kathleen Kimball, *The Music Lover's Quotation Book*

It's a good thing to be able to like any kind of music, but it's also wise to be cautious that enthusiasm for any one type of music—jazz, pop, rap, or Ludwig van Beethoven—does not prevent the listener from hearing the beauties to be found in other kinds of music.

In free-form improvisation, beauty—let's call it an appreciation—is to be found in the ways the performers relate to each other as they make the music. The music itself is simply the vehicle that enables this interaction. The objective of this kind of music making is not the creation of Beethoven symphonies. It is to reveal the participants—performers and listeners—to themselves.

A basic premise of free-form improvisation is that beauty and meaning are to be found in whatever music happens. The performers' choices of harmony or dissonance, of meter or disjunct rhythm, of abstraction or representation, will be meaningful in and of themselves—regardless of what was specifically chosen. Their choices reveal the personal values and sensibilities of the performers to themselves and to others. Any perceived presence or absence of beauty or understanding is as revealing about the performers or listeners themselves, as it is about the music.

The CMM facilitator, upon hearing a participant indicate that he or she did not like an improvisation or a particular aspect of it, may ask, "What was it about the music that you didn't like?" or "What does that tell you about the music?" and "Does that tell you anything about yourself?" Such questions also apply to any statement by a participant who shows enthusiasm for the music.

The process of creative music making seeks to open minds—to expand the ability of participants to understand and appreciate more about the music that they and others create, and perhaps to extend that appreciation into other aspects of their lives.

7

CREATIVE MUSIC MAKING

I love music passionately, and because I love it I try to free it from barren traditions that stifle it. It is a free art, gushing forth—an open-air art, an art boundless as the elements, the wind, the sky, the sea. It must never be shut in and become an academic art.

—Claude Debussy, quoted in Nat Shapiro, *An Encyclopedia of Quotations about Music*

Creative music making sessions consist of four simple, interrelated steps: playing, recording, listening, and questioning. All four steps are equal in importance.

CMM incorporates free-form improvisation on musical instruments or with the voice, but it also provides a way to explore and understand the music by including active listening and questioning as integral elements in addition to active playing.

Because CMM normally develops in progressive stages over time, there are two recommendations for getting started:

1. Have an experienced facilitator at CMM sessions, at least in the beginning
2. Plan a regular time each day, week, or month for CMM sessions

HAVE AN EXPERIENCED FACILITATOR AT CMM SESSIONS

Adapting [an] effective way of teaching improvisation ... to teaching in a classroom raises many problems: ... avoiding the establishment of a set of generalized rules and always allowing an individual approach to develop. ... And the only places where, to my knowledge, improvisation is successfully taught in the classroom is in those classes conducted by practicing improvisers.

—Derek Bailey, *Improvisation: Its Nature and Practice in Music*

Especially in the first few CMM sessions, the role of the facilitator—to support all of the session participants in their process of developing self-directed playing and listening skills—is extremely important. The facilitator will be actively engaged as the primary source of direction in the beginning sessions.

In the playing step, especially in the beginning phase, the facilitator is responsible for providing a model that the participants can emulate in their own playing.

In the first session, right at the very beginning, the facilitator should describe the four steps of CMM. Second, the two rules for playing free-form improvisation should be described. The facilitator may even have the participants repeat the two rules out loud together before they play.

Then, the surest way for a CMM facilitator to gain the immediate trust of participants is to start by selecting one of them and playing in a recorded improvisation with that person right away. This action will not only model the facilitator's musicianship, but also her or his openness in receiving and accepting the participant's contribution to the music, whatever it may be. In so doing, the facilitator and participant can be seen and heard by all participants to be equally vulnerable, which is a basic condition in the normal musical risk taking that CMM seeks to encourage.

The facilitator's ability to play with confidence and to support whatever music the first-time improviser creates is a crucial element in insuring a positive initial experience with CMM. It insures that the introductory improvisation will benefit from the facilitator's experience in creating free-form music that is engaging for the listeners and is "organic"—in other words, alive.

After listening to a playback of the recorded improvisation, the facilitator, as an experienced improviser, must draw upon that experience in determining the most appropriate question with which to start the discussion. In the questioning step, the CMM facilitator then actively guides participants in raising their own questions at appropriate times within the context of the discussion as it progresses. If the participants are unable to raise questions on their own, the facilitator assumes the responsibility for raising relevant questions him/herself.

As a general principle, the facilitator may assume that every participant has an individual response for each question raised. As with the playing step, the object is not to elicit right or wrong answers but to elicit individual perspectives for comparison with other differing perspectives, which the facilitator should encourage to be expressed. A good rule of thumb is that the main difference between the facilitator and new CMM participants is not that the participants don't know how to respond to the questions, but that they don't know which questions to ask. Once a question is raised, the responses normally flow quite readily.

The facilitator insures that all four of the CMM steps (presented in more detail later) are present—particularly the listening step—and that the two rules of free-form improvisation (also given again later) will be applied. It is important to guard against impatience and the temptation to short-cut the process by omitting one or more of the four steps. Listening and questioning are at the core of CMM. Without the support of an experienced facilitator, beginning CMM participants may be overly eager to focus on the active playing step at the expense of the active listening and questioning steps.

Another responsibility of the facilitator is to maintain an awareness of the similarities and differences among successive improvisations as they follow one another in a CMM session. Normally, there will be a tendency for beginning participants to mimic preceding improvisational pieces in length, mood, form, or style, especially in the initial session, immediately following the facilitator's participation in the first improvisation. The first few improvised pieces may be strongly influenced by the initial piece and may even fall into a pattern that is continually revisited with each piece that follows. The participants might find themselves in a musical rut that may provide them with a sense of security but also may stifle other musical options. It may simply be easier for participants to imitate what has happened in earlier improvisations than to take risks that lead into other musical directions.

If this pattern occurs, the facilitator may pose questions that bring it to the attention of the participants: "Did this piece sound like the last piece? Why do you think so?" The facilitator may also raise questions that lead participants in new directions: "Are there any other alternatives? What other musical directions are possible?"

Another important task of the CMM facilitator is to be on the lookout for body tension in any of the participants as they are playing and to encourage them to relax. This can be done easily by simply asking the participant, "Are you aware of any tension in your body as you play? Do you feel relaxed?"

Gradually the role of the facilitator in directing activities in CMM sessions should diminish. As participants' experience in CMM increases, the facilitator must also develop a sense of when to abstain from intervening in the process of playing, listening, and questioning in order to allow, as much as possible, the participants to work through the issues they raise on their own. It is desirable that as participants gain in CMM experience, the facilitator's involvement diminishes, until eventually it is the participants themselves who have the motivation and skills to drive the process and share in the facilitator's role.

PLAN REGULAR CMM SESSIONS

I never practice, I always play.

—Wanda Landowska, quoted in Kathleen Kimball, *The Music Lover's Quotation Book*

As in the development of any skill, it is important to have regular opportunities to continue learning and to apply what has been learned.

Perhaps the biggest difference between "practice" as it is commonly done and "practice" as it is recommended here is that in CMM, practice occurs in an ensemble environment, and not in solitary isolation. In CMM, the group provides support and encouragement, not competition. The learning process is lateral and peer driven, not vertical and authority driven.

Also, as will be explored below, in CMM the goal is to develop musicianship through self-directed questions and self-searching answers rather than through answers provided by authorities. As the ability to raise and resolve pertinent questions is exercised in a supportive group environment, the possibility of improving the quality of practice in any other musical genre increases, even if practicing alone.

CMM AS A COURSE OF STUDY

Mastery comes from practice; practice comes from playful compulsive experimentation … and from a sense of wonder. … In practice, work is play, intrinsically rewarding.

—Stephen Nachmanovich, *Free Play: Improvisation in Life and Art*

Regular CMM sessions will be most effective if they occur within a formal course of study, with one or more sessions per week over a set number of weeks or months during a school semester. This structure would be particularly appropriate for career-track music students in a conservatory of music or a university music department. It is the best method for insuring that all of the playing phases of CMM can be experienced by participants. These playing phases are described in more detail on the following pages.

CMM weekly or biweekly course sessions need not involve large classes. In fact, by keeping a session small—involving only one, two, or three students plus the facilitator—it becomes more likely that individualized attention will be received by participants. Additionally, there can be an increased amount of flexibility in scheduling CMM sessions if fewer participants are involved. Another possibility might be to alternate weeks between large-class CMM sessions and small individualized sessions.

The main obstacles are likely to be finding the times and places to conduct CMM sessions. Some institutions might have fewer rooms—or even no rooms—available at certain times. Simple recording and playback equipment would also have to be available.

The facilitator's available time could conceivably be limited as well, but after the first few CMM sessions, participants ought to be have enough experience to conduct sessions on their own from time to time without a facilitator present.

Regular monthly or bimonthly CMM sessions can be of value to student and professional ensembles by providing opportunities for the ensemble members to interact in a more relaxed and playful musical environment. This kind of playful interaction among ensemble members can be beneficially absorbed into the other musical activities of the ensemble.

CMM AS A WORKSHOP

Another option is to present CMM in the form of a workshop, with three or four sessions, ideally over several days. This may be the best option for professional musicians or others having limited time available to be away from their other commitments. The main concern is that enough structure and time be provided early on to enable beginning participants to progress through the phases of development in performing free-form improvisations.

Either structure—study course or workshop—can provide a firm enough groundwork to achieve desirable outcomes such as the motivation to conduct self-directed CMM sessions with friends and colleagues, the confidence to schedule improvisations on recitals and concerts, or the desire to apply CMM processes in types of music other than free-form improvisation.

8

CMM STEP 1—PLAYING

We learn to improvise by improvising.

—John Wyre, *Touched by Sound*

For all performers in the playing step of a creative music making session there are only two rules:

Rule 1: Performers may play (or not play) anything they wish on any available instrument of their own choosing—there are no mistakes.
Rule 2: Performers should listen as deeply as possible to themselves and to the other performers, but it's important that it be perfectly clear to all participants—players and listeners alike—that there is no penalty for breaking this rule.

No penalty? Isn't there always a negative result in a musical experience if listening is neglected?

The point here is to avoid placing a burden of expectation on the performers and listeners that can stifle imagination and risk taking. In the discussion that will occur in the questioning step of CMM, it will be quite apparent that most participants know who was listening and who was not. It will also be clear to all that not listening has consequences (as opposed to penalties), and that the consequences may be (but are not always) negative. This lesson can be learned much more deeply through self-discovery without the stress imposed by concern about penalties. Establishing a trust among all participants that they can truly be free is more important in the long run than any short-term slap on the wrist.

BEGINNING ENSEMBLE SIZE

It is important that participants, in their first experiences with CMM, have an environment conducive to open expression. It is recommended that for CMM participants playing in their first improvisation the ensemble size be limited to two players, with one of the

Fig. 8.1 It is recommended that for CMM participants playing in their first improvisation, the ensemble size be limited to two players, with one of the performers, ideally the facilitator, already possessing in-depth experience in freeform improvisation.

performers, ideally the facilitator, already possessing an in-depth experience in free-form improvisation.

The reason for this may not be apparent, but having a larger ensemble at this stage would be like placing someone in a room full of strangers and asking that person to have a meaningful conversation with everyone at once. The more people in the room, the more complex the challenge; the room would seem to be filled with unintelligible chatter. It would be a challenge to listen carefully enough to hear, let alone understand, everyone else's ideas. It would be even more difficult, if not impossible, to present an idea and have any sense that it's being perceived by everyone else.

Admittedly, such a situation could be interesting, too. In fact, it is not unheard of for some improvisation classes to be structured precisely in this way. But in a one-on-one environment, especially with an experienced improviser, the beginning participant's involvement in the musical give-and-take is much more likely to be positive and meaningful.

It might seem that having the freedom to play whatever one wants would lead to some sort of uninhibited frenzy. In fact, the opposite usually occurs. There are typically three phases—not necessarily occurring in sequence—of development in performing free-form improvisations.

PLAYING PHASE 1

Beginning participants in their first playing sessions of CMM are likely to be overly cautious, wary of risk taking, skeptical about the concept of "no mistakes," and fearful of being judged negatively by other participants. Every attempt to make a sound can become a personal trial; every action is assumed to be judged by others. It can be very difficult to let go of existing concepts about what music "should" be. Every sound becomes a test of the concept that "there are no mistakes."

While some participants may handle such concerns with ease, or even with indifference, great care will still have to be exercised by the facilitator in creating a supportive and non-threatening environment for first-time improvisers. For some of them, it might be an enormous challenge to take responsibility for making the performance decisions that are necessary in improvisation.

As an example of the kind of reaction that can be presented by a first-time improviser, in one CMM workshop the facilitator selected a participant to play in a session-opening improvised duet. After conveying the two playing rules, the facilitator began the improvisation by simply playing a sustained tone on a single pitch. The tone continued, with breaths, for almost an entire minute—an exceedingly long time—while the participant, gradually realizing that a response was being solicited, became frozen with a "deer-in-the-headlights" kind of wide-eyed stare. The first-time participant simply could not decide what to play. Making a choice from among the overwhelming number of possibilities was much too risky.

What were the risks? Since trust had not yet been established with the facilitator, the participant could not be sure whether or not there were any hidden expectations. The risk was in the possibility of embarrassment in front of the facilitator and the other participants, should the player fail to rise to any unspoken expectations: What if I don't do exactly what the facilitator wants me to do? How will I be judged?

To make matters even more dire, it soon became apparent to the participant that the substantial length of time that had passed without any playing also created risks for embarrassment. Upon realizing that the participant was facing such a dilemma, the facilitator stopped playing the long tone. With a smile the facilitator complimented the first-time improviser for following Rule 2 by being a great listener. A few reassuring remarks about Rule 1 were offered, along with the reminder that the improvisation was just going to be for fun.

The facilitator then played another long tone on a different pitch as the first-time participant tentatively played one single short note, then a couple of notes, and eventually entire phrases. Every time the participant played, the facilitator responded musically in some way to confirm the participant's musical choices. The outcome for the participant, especially after hearing the playback of the improvised piece, was one of amazement and joy that the music was actually interesting to listen to, and of relief that no embarrassment had been suffered.

After all, the goal of improvisation in CMM is not to make perfect pieces of music beyond criticism (Ludwig van Beethoven symphonies) or to render groundbreaking performances (Miles Davis solos). The immediate goal is to make whatever music is made simply to see what happens, and then to examine it and try to learn from it.

PLAYING PHASE 2

When it is accepted that the freedom to do whatever one wants really exists, based on having actually experienced it in prior free-form improvisations, a second phase of development exists, in which performers consciously take more risks, testing the limits of their newfound musical freedom.

The experiments undertaken by participants may involve severe departures from normal music making. CMM participants in this phase have made sounds in their improvisations by dropping things on the floor, knocking furniture around, making wild vocalizations, clip-clopping their shoes, squeaking their sneakers on the floor, banging or tapping on the walls, and even scratching the recording microphone.

As long as no property is being damaged (no examples of such damage come to mind) this kind of playful experimentation should be neither discouraged nor encouraged by the facilitator. Rather, the facilitator can refer to the actions of the participants in the questioning step: "What were you trying to do? Did the sound(s) produce the intended effect in the music? How did the sound affect the other players? After hearing the playback, do you think the experiment worked?"

In this phase, performers may be so focused on their own musical experiments that they are entirely oblivious to what other performers are doing. Upon hearing an immediate playback of a recorded CMM session, performers in this phase might have little recollection of having played what they are hearing. The thing that is still missing is the balance provided by good listening while playing.

Again, playing one-on-one with an experienced improviser is recommended as the most supportive learning environment at this stage, for the same reasons given above for playing Phase 1. The beginner's listening skills are still being developed.

PLAYING PHASE 3

With just a little more experience, a third phase of development is reached, in which listening and playing come into more of a balance. The performer's task is now to be connected to the music and to what the other performers are doing. Whether the connection is harmonious or starkly contrasting is not as important as whether or not the connection derives from listening. When this stage is reached, performers are primed to deepen their musicianship—by making spontaneous, intuitive, as well as logical performance decisions; by taking risks in presenting musical ideas; by determining which musical ideas to embrace or discard; and by relating to other performers in a common goal, to make interesting music.

As the number of players is increased in a free-form improvisation, the musical environment becomes more complex and listening becomes more important. So, only when beginning performers demonstrate that they are able to actively listen should ensemble size begin to expand, perhaps to three or four players at first, and later to even larger groups.

As players develop their ability to listen to "the music"—an organic whole, comprising the combined music making of all players—the inward search for appropriate musical responses intensifies. The task at hand becomes how to contribute to the music.

It is in this third playing phase that the real benefits of CMM begin to emerge for the participants, and the facilitator's presence will be critical in reassuring them through the earlier playing phases. CMM as a pedagogy for improvisation is a process of growth that occurs over time, and it is important to insure that the process is not mistakenly cut short.

Beginning participants at all age levels may be tempted, after experiencing their first improvisations, to think that because the rules are so simple they have already done everything possible and that further participation will only consist of revisiting what they have already done. They may say to themselves, "Is that all there is to free-form improvisation? I can do that! Why do I need to spend any more time on it?"

The participant asking such questions is still unaware of the full range of playing possibilities and the potential for musical growth. Participants in the early playing phases have not yet experienced improvised pieces that are complex and contain musical ideas that are more fully developed. It is only natural that in the first two playing phases relatively simple pieces will unfold using simple musical ideas. However, simplicity in content can still provide ample musical material for listening and discussion.

The facilitator knows that such results are only the first small steps in the process of reaching for greater heights. It is the facilitator's role to continually find ways to help participants to see new possibilities in tapping into and expanding upon their playing vocabularies. One way for this to be accomplished is through raising appropriate questions for discussion in the questioning step of CMM. Another way is for the facilitator to join in the playing and to introduce musical ideas that lead into new areas of exploration, while carefully refraining from overwhelming the ensemble. The facilitator provides the guidance that insures that the process of musical growth for the participants progresses through all of the playing phases.

Fig. 8.2 With experience, listening and playing come into balance. At this stage participants can improvise one-on-one together, without the facilitator playing.

HOW IMPORTANT IS IT TO HAVE A (RHYTHMIC) GROOVE WHILE PLAYING?

The concept of time is irrelevant when you're playing time

—NEXUS member Michael Craden, at York University, Toronto, Ontario, 1974

Unlike most styles of popular music or Jazz, in free-form improvisation it is not essential to have a rhythmic groove—also referred to by musicians as "time"—although there should be nothing to prohibit it if a player wants to play one. Just as with the other elements of music—melody, harmony, tone, and the like—rhythm and meter are all determined by each individual player during performance. The music may at times be completely without meter or groove. At other times there may be a single groove or multiple contrasting grooves. It's entirely up to the players.

A constant rhythmic groove is something that is present in almost all of the music that is in our environment, from the radio in the morning, to the elevator during the day, to the CD player at night. Since each individual's music making in an improvisation is influenced by that person's musical vocabulary, it is only reasonable to expect that the rhythmic grooves that are in the environment—in pop, rock, Latin, and jazz styles—will be a part of that vocabulary and will emerge in the improvisations.

As a consequence, it is much more likely that an improvisation in a CMM session in North America will have a 4/4 or 2/4 groove than a 7/8 or 9/8 groove, unless the vocabulary of the participants includes experiences in music from other environments—say, Africa, or perhaps Greece.

The facilitator can be a source of vocabulary expansion for the participants, simply by playing with them in an improvisation and introducing more exotic time grooves.

A regularly repeating rhythmic pattern, technically called an *ostinato*, is one tool that can be used by players to bring a sense of unity to an ensemble's improvised music. It can also become a security crutch that stifles listening and overwhelms other musical ideas. In the event that this is observed, it is the responsibility of the facilitator to insure that it is discussed (without judgment) in the questioning step: "Was everyone comfortable with the groove? Did anyone become tired of the groove? Did anyone try to change the groove?"

The facilitator might also ask, "To what extent were you thinking about time? Did thinking about the groove take your mind away from hearing what the others players were doing?"

HOW DO I KEEP IN TOUCH WITH THE MUSIC WHILE PLAYING?

Musical performance is born in those same sublime regions from which music itself has descended. Whenever the music is in danger of becoming earthbound, the performer must elevate it and help it to regain its original ethereal quality.

—Ferruccio Busoni, quoted in Ian Crofton and Donald Fraser, *A Dictionary of Musical Quotations*

At any given moment in a free-form improvisation, each player has at least three possible performance options to choose from:

1. Listen and try to play something similar to what is heard; match another player's rhythm, pitch, tone-color, and so on; blend in with the overall mood
2. Listen and try to play something different from what is heard; change or make a contrast with another player; alter the mood
3. Listen and be silent; don't play at all

Knowledge that these three options exist can be acquired directly from the process of questioning and answering guided by the facilitator. For example, the facilitator might simply ask the participants, "What options do you have in responding to what you hear while playing?"

Even with very experienced musicians, moments of disengagement from the music may occur—possibly the result of unrelated thoughts, daydreaming, or even boredom. Of course, moments of disengagement from the music can occur in any kind of music making, regardless of the musical genre or style. In free-form improvisation, when a player realizes that he or she is disengaged from the music, the same three response options are available: listen and play something similar; listen and play something different; or listen and stop playing.

There may be other possible responses. For example, a player may choose to continue playing in a disengaged way by not consciously relating at all to what is going on in the music. It's important to mention that this response is acceptable, too. In fact, such a response can provide a source to draw upon for discussion in the questioning step of the CMM session.

HOW DOES THE MUSIC END IN A FREE-FORM IMPROVISATION?

The music comes to an end when all of the players have individually decided to stop playing. Any player wanting to continue playing after others have stopped is free to do so. In fact, the urge to continue on with an idea may serve as the key that opens up a deeper or entirely new avenue of musical content.

In the event that one player continues to play alone long after the other participants have stopped playing, the facilitator should insure that the player is permitted to continue until he/she decides to stop. Undoubtedly, such a performance will be a topic in the group discussion later.

ARE THE PLAYERS ALLOWED TO MAKE UP ANY RULES OR LIMITATIONS ON THEIR OWN?

> *[A] teacher is well advised to be quiet from time to time about even the most ordinary facts, so that students may have the freedom to make those facts their own.*
>
> —Paul Woodruff, *Reverence*

Each player in a free-form improvisation is free at any time to play under self-imposed rules. On occasion, though, it can be fun to make up shared rules as a means of structuring a free-form improvisation. While any made-up rules or limitations will probably not be harmful to the musical growth that CMM seeks to foster, caution should be exercised that the made-up structures do not inhibit the freedom of the players to listen and play whatever their intuition tells them to play. The temptation to change a free-form improvisation into a quasi-composition—that is to say, to reduce risk by changing an unfamiliar musical entity into a more familiar one, or to manufacture and cling to the security of rules or externally imposed structures—can be very strong.

However, this temptation ought to be resisted for the purposes of CMM, especially with novice participants, because externally imposed rules can cause the players to focus on the externally imposed ideas instead of on listening to what's actually happening in the music.

That being said, supplemental playing rules may be used effectively as exercises if they focus specifically on listening while playing and making intuitive responses. Such exercises can be study tools. Temporary, artificial rules may be employed, as long as the exercises are not mistakenly confused with free-form music making itself.

IS IMITATION ALLOWED?

Every artist has to learn how to copy before he can invent.

—Laurence Boldt, *Zen and the Art of Making a Living*

Imitative playing—copying one or more of the musical aspects in the playing of another ensemble member (pitch, rhythm, dynamics, tone color, and so on)—is a very typical response in the first playing phase, and it should not be discouraged. Imitative playing also continues into all later phases. However, as experience increases and as listening skills improve the imitation response tends to become more refined and less direct. As a player's confidence in his or her own intuitive reactions increases, the imitation becomes more and more subject to the influence of that player's own creativity. The line between imitative playing and intuitive playing becomes increasingly blurred, and an overall sense of unity and wholeness in the music becomes more apparent.

One characteristic of this unity in the music is the presence of what may be called *consonance* or *synchronicity.* Consonance may be defined as a moment in the music when communication among the players is at a peak. This may be evidenced in the music by the presence of a simultaneous tonality, rhythm, or any other musical element to which the individual player's seem to have completely surrendered themselves.

It is difficult, if not impossible, to know to what extent the occurrence of such a consonance is brought about either by simple coincidence or by deep listening. However, it is generally true that while brief consonances may occur in the earliest stages of CMM, more experience leads to an increased number and length of such consonances. This is a deeper level of music making—a greater sense of connectedness to one's self and to the other players. The search for consonance is, in short, what CMM is primarily about.

SELF-IMITATION AS A CMM EXERCISE

One good CMM exercise involves self-imitation. First, a selected participant improvises a single phrase, melody, or motif. Then, that same player tries to repeat it, as exactly as possible. Next, the participant tries to play it over and over, each time trying to vary it in some subtle way—in pitch, dynamics, rhythm, or speed, for instance—until ideas are exhausted. One by one, each succeeding participant around the circle does likewise. The goal of this exercise is to provide the participants with experience in developing their own musical ideas.

IMITATING ANOTHER PLAYER AS AN EXERCISE

A second CMM exercise is a variation of the above. A selected participant begins by making up a single phrase, melody, or motif. Then, one by one, each succeeding person in the circle tries to repeat it, as exactly as possible. Finally, one by one, each succeeding person in the circle tries to vary it, continuing around the circle as long as interest holds. The facilitator can guide the exercise simply by prodding each successive player with the question, "What else can you do?"

MAKING A CONTRAST AS AN EXERCISE

In another variation, each succeeding participant in the circle tries to play something completely different from the preceding person's idea, while still trying to make some musical connection.

A CLAP/CHANGE EXERCISE

Another musical exercise is that upon a predetermined signal—a hand clap, or the sounding of a particular instrument—all performers will make a change to something completely different from whatever they have been playing. The signal may be given only once or frequently. It may be given by the facilitator or by one of the players or by each player in turn.

TAG-TEAM IMPROVISATION

A select group of players begins the improvisation; all remaining participants are standing (or seated) together nearby. Whenever a player determines that the time is right, that player stops playing and approaches the standby group. Anyone in the standby group who has not yet played in the improvisation raises his/her hand. The retiring player selects one of the standbys to join in the improvisation.

GROUP TAG-TEAM IMPROVISATION

All participants are divided into three or four groups. Group 1 begins the improvisation, and upon a predetermined signal given by the facilitator or one of the players, Group 2 joins in the improvisation as Group 1 fades out. The process is repeated for each of the remaining groups. The last group ends the improvisation. Another possibility is that after the last group plays, a signal is given and all groups play together until the music ends.

PRACTICING CMM ALONE AS A SOLOIST

It is certainly possible for anyone to play, record, listen to, and raise self-directed questions about a solo improvisation. However, without the support of either coparticipants or a facilitator, it will take a level of understanding attained by enough prior experience in improvisation to make the music meaningful. Without the ability to rely on the input of musical ideas generated by other players, all of the musical ideas must be created and developed single-handedly—a difficult, if not impossible task for beginning improvisers. Therefore, solo improvisations are only recommended for those players who already have a substantial amount of experience in improvisation.

DRUM CIRCLES

I got rhythm, I got music.

—George Gershwin, "I Got Rhythm"

The four-step process of CMM can also be applied to drum circles, a form of musical expression that can be a lot of fun and that is now practiced all over the world. Obviously, due to the nature of drums the main element in drum circle music is rhythm, to the virtual exclusion of melody and harmony.

Typically, participants in a drum circle focus on maintaining a rhythmic groove that has been established either by the master drummer or by one of the other participants. The rhythmic groove is normally a fixed repeating pattern or ostinato that is played by the participating drummers with a certain "time feel" derived from idiomatic music—for example, from jazz or ethnic styles of Africa, Cuba, Brazil, or Japan.

Participants in drum circles usually play on traditional drums (or commercially manufactured copies of traditional drums) from the ethnic cultures from which the rhythmic grooves are derived. Idiomatic drumming techniques are learned, thereby building a repertoire of differing drum strokes and sounds. These are combined to form the entire musical vocabulary from which a "feel" is created. When the feel or groove is firmly established by the ensemble, opportunities are usually provided for individual participants to improvise a solo over the ensemble, drawing also on the acquired vocabulary of rhythms, strokes, and sounds.

Improvisation in most drum circles can be categorized as idiomatic, and as such it is somewhat different from the free-form improvisation that occurs in CMM. Nevertheless, the experience of playing music in a drum circle can be enhanced by the additional CMM steps of recording the music, listening to it played back. and discussing it.

What drum circles and CMM have in common is intuitive music making in a supportive ensemble environment with a minimum of externally imposed stress and a maximum of involvement.

9

CMM STEP 2—RECORDING

I have music here in a box ... ready at a touch to break out of its prison.

—Aldous Huxley, *Music at Night*

In creative music making sessions, recording goes hand in hand with playing. The recording step does not actually occur separately from the playing step; both occur simultaneously. But the recording step is significant enough to be considered separately, because it alone provides the means for players and CMM participant listeners to hear their own performance and consider it in a nonplaying context. It enables participants to reflect on what they have created, and to learn about themselves through self-analysis. It enables players to hear themselves as others hear them, providing a context for the players to better understand the observation and analysis of others. It enables listeners to hear the music without the influence of the visual cues they receive when they watch the performers play.

Making a recording can also give the playing a heightened sense of importance for the performers, because the recording becomes a concrete documentation of the improvised music, which would otherwise exist only in the individual memories of the participants.

It is certainly true that a high quality sound system would be better than one of lesser quality. It would simply make for better listening. However, this must be balanced by the practical reality that a high quality recording and playback system might not always be available. If that is the case, it will be perfectly acceptable to use whatever type of recording and playback system is available. For the purposes of CMM even a poor quality sound system will provide enough information for participants to benefit from listening to the playback of their performance, particularly with it still fresh in their minds.

It will be best to keep the work of making recordings as simple as possible in order to minimize time-consuming technical problems. Only a minimal amount of recording equipment is necessary:

- A digital audiotape (DAT), mini disc, or audiocassette recorder/player
- One microphone (monaural or stereo)
- An amplifier (monaural or stereo); or an integrated amplifier like a guitar amplifier

- One (or two) playback speaker(s) if an integrated amplifier is unavailable
- Patch cords (microphone to recorder; recorder to amplifier; amplifier to speaker[s])
- An AC electrical outlet (with surge-protected extension cord)

Ideally, the equipment should be entirely set up and ready for use by the posted starting time of each CMM session. For the best use of time it is recommended that one person other than the facilitator be assigned the sole responsibility for operating the recording equipment—setting up the equipment, loading the media (DAT, mini disc, or audiotape cassette), recording CMM performances at proper volume levels, locating the correct media time addresses for playback, and running the playback at an appropriate volume level.

Another participant might be assigned the responsibility of logging the names of the performers for each improvisation on a sheet of paper for reference in the future.

It's also a good idea to begin each recorded improvisation with a spoken announcement of the players' names. The date and place of the CMM session can also be mentioned for additional reference. Following the announcements, a few seconds of silence will help to clear the air prior to beginning the playing/recording.

In a CMM session, the playback of each recorded performance can occur either immediately following the end of the performance, or following a postperformance discussion, depending on the wishes of either the facilitator or the participants.

RECORDING FOR EDUCATIONAL PURPOSES

In most situations the recordings made for playback in CMM sessions can be discarded or erased after they have been heard and have served their purpose as an educational tool. They might also be kept on file for reference or for listening at some future CMM session, in which case participants could have an opportunity to hear their music again with the added advantage of much more listening and playing experience.

RECORDING FOR PROMOTIONAL PURPOSES

If high quality recording equipment is available, or perhaps even if a recording studio with a professional sound engineer is possible, it might be worth considering professional CD-quality recordings of CMM improvisations, whether they occur in regular sessions or in gala concerts. Such recordings could be assembled to make a CD master in order to give reference copies to the participants or even to make a commercial or promotional product. Any person or institution responsible for promotional or commercial uses of CMM recordings should also be aware that such uses could involve some important legal issues, such as copyrights or the license to use participants' performances.

Any recorded music can be copyrighted, whether or not it is in written form. Generally, in the case of free-form improvisations it should be assumed that the players in any given piece are the cocomposers and copyright owners, whether or not a copyright form has been formally registered with the government copyright office.

In addition, the person or institution that organizes and assembles the improvised pieces onto a CD should be assumed to be the copyright holder of the CD compilation. Just to clarify, a CD compilation copyright is different from the copyrights for the individual compositions, which are held by the composers or their publishers. It would be a good idea (and, depending on the specific circumstances, maybe even necessary) for the

Fig. 9.1 It is perfectly acceptable to use whatever type of recording and playback system is available. In this case a Minidisc player, stereo mic, and integrated guitar amp (mono amp with EQ and speaker) were operated by the facilitator.

copyright holder of the CD compilation to obtain written permission from the performers to use their performances, and to indicate to the players in writing the intended use of the CD. The recording might, for example, be used to promote an institution or persons or ensembles in an institution.

Realistically however, in most cases, dealing with copyright issues would merely serve as a courtesy to the participants, because there will probably be little or no money generated for any of the copyright holders. But in the case of an institution or ensemble seeking to use the CD as a fund-raising device, it would be wise to consider copyright matters.

Copyright law can be very complex, and it is not the intent in raising copyright concerns here to delve into this topic in any detail. Suffice it to say that there are a number of readily available publications that deal with the subject of copyrights that can be informative and helpful, before seeking professional legal advice. Consumer-oriented books dealing with music copyright issues can be found in public libraries and even in large bookstores.

RECORDING FOR COMMERCIAL PURPOSES

If the intention is to make commercial use of a recording—that is to say, to make money from producing a CD—then there are a number of other concerns, especially if any of the participants are members of a musicians' union. For the purposes of CMM, the main point to be made is that the commercial use of session recordings is certainly one possible way for participants to take advantage of their CMM experiences, although it would also be very wise to have a practical understanding not only of copyright matters, but also of the potential market for any such recording.

TRANSCRIPTION FROM RECORDINGS

Another possible way for participants to take advantage of recorded CMM improvisations is to transcribe from them, in whole or in part, to create written compositions that can be used privately or copyrighted and published. Depending on what material is transcribed or who the players are in any particular piece, the same issues of courtesy and copyrights would have to be addressed.

It would certainly be possible for one, two, or more players to play and record a free-form improvisation solely for the purpose of transcribing and publishing a cocomposed piece. This is actually how some professional composers work. Again, the point here is that making a transcription from a CMM improvisation is simply another possibility for CMM participants.

10

CMM STEP 3—LISTENING

The word improvisation is commonly, yet erroneously, associated with a specialized meaning in music—that which relates to performing music without reference to notation. Actually, improvisation occurs when we audiate music, when we listen to music being performed by others, when we read notation, as well as when we perform music without the aid of notation.

—Edwin E. Gordon, *Rhythm*

Good listening (audiating) is the key to well-rounded musicianship. In a cultural environment where physically active "doing" is valued highly, it's sometimes necessary to draw attention to mentally active "doing," which is what good listening is.

In creative music making, listening goes hand in hand with playing. Good listening improves the ability to perceive nuance, and to reach a greater depth in intuitive, nonlinear modes of understanding. It is the primary means by which musical concepts, familiar and unfamiliar, can be assimilated, stored, and later put into use in performance.

In performing free-form improvisations, the musical ideas that have been internalized through past listening and performance experiences are the reservoir from which performance ideas may spring. This seems obvious. It is very unlikely that a performer will play a 12/8 rhythmic ostinato (common in African, Afro-Cuban, and Korean music) or a six-tone scale (common in Balinese music, compositions by Claude Debussy) in an improvisation unless that performer has already heard them in music and internalized them in his or her musical memory.

It is this internal store of musical ideas that, essentially, is each musician's vocabulary. It should also be obvious that the larger a musician's vocabulary is, the greater the range of possibilities in that musician's music making. Building musical vocabulary is a lifelong exercise.

Good listening is the best method of expanding one's musical vocabulary. It is the gateway through which new and unfamiliar musical ideas can be received.

Fig. 10.1 Good listening is the key to well-rounded musicianship and it goes hand in hand with playing. It improves the ability to perceive nuance, and to reach a greater depth in intuitive, nonlinear modes of understanding.

LISTENING TO THE PLAYBACK OF A CMM IMPROVISATION

> *I remember the amazement in realizing "the more you listen the more you hear," the delight in registering sounds that had always been present but I had never heard, the ecstasy of knowing this is a lifelong experience, infinitely expandable, basically musical.*
>
> —W. A. Mathieu, *The Listening Book: Discovering Your Own Music*

Basic listening techniques can be easily learned, and once assimilated, they can lead to good musicianship. In the first CMM session, before listening to the playback of recorded free-form improvisations, the facilitator might ask participants to offer their own thoughts on good listening techniques. A few possibilities are:

- Relax (with a deep breath, body tension released, mind cleared). A good exercise to prepare for listening is to stand with arms at the side. After taking a few unforced deep breaths, gradually relax the body, first by rolling the head gently, and then by rolling the shoulders. Next, concentrate on relaxing the muscles of the arm, wrist, hand, and fingers. Finally, relax the muscles in the torso, hips, legs, and feet. Breathe slowly and deeply for a short while.
- Do not rush to judgment about what is heard (what it is/what it means). Making judgments about what is heard (i.e., good/bad, awful/great, etc.) is not helpful in good listening. In fact, rushing to judgment can inhibit one's ability to hear clearly and to pay attention to details.
- Try to notice as much as possible about the sounds that are heard.

OBJECTIVE AND SUBJECTIVE LISTENING

Over time, I began to feel that the organizing principle of the universe is "relatedness," and that this is more fundamental than "thingness."

—Joseph Jaworski, *Synchronicity*

There are two practical ways of noticing what is heard. The first is objective listening. Try to identify things such as the various objective elements of the music—instrumentation, melody, pitch, volume, rhythm, tempo, harmony, mode, timbre, form, motif, counterpoint, and the like. In objective listening the listener identifies those concrete things in the music that every other listener can generally hear and agree upon.

Subjective listening involves noticing the feelings or moods that the music evokes. In subjective listening the listener looks inward and identifies a very personal mood or emotion—the effect of the music upon the listener. While the emotional effect may be described in the same way by other listeners, it is an interpretation rather than a concrete quality of the music. Some listeners, if not familiar with a particular kind of music, may have a very different emotional response to it.

Using both modes of listening, participants should also try to notice any changes that occur in the music over time.

In CMM sessions, particularly in the first few, where the playing is one-on-one, all of the nonplaying participants will be observers/listeners. In listening to the playback of recorded performances, everyone—players and nonplayers—will be actively involved as listeners.

BUILDING VOCABULARY THROUGH SUPPLEMENTAL LISTENING

With recording, the music of the world becomes available at any moment—just like an encyclopedia. … Music becomes plural.

—Marshal McLuhan, quoted in Nat Shapiro, *An Encyclopedia of Quotations about Music*

In addition to listening to the playback of improvisations, another kind of listening in CMM sessions is aimed at building the musical vocabulary of participants. In CMM this is called, for lack of a better term, "supplemental listening."

Time can be made in any CMM session to devote to supplemental listening, particularly if a topic under discussion can be better illuminated by listening to a musical example. Certain CMM sessions may be entirely devoted to supplemental listening, or perhaps each session could either begin or end with it.

The facilitator can select the recorded music to be heard, or the participants can each be asked to bring recorded music to a CMM session for listening. Participants should be encouraged to select music they really like, as well as music they think is capable of expanding the vocabulary of other participants.

Supplemental listening in a CMM session might begin by listening to relatively familiar recorded music—pop/rock (e.g., Elvis Costello, Elton John); popular American song (e.g., Leonard Bernstein, George Gershwin); jazz (e.g., Duke Ellington, Branford Marsalis, Red

Norvo); and classical (e.g., Aaron Copland's "El Salon Mexico," Wolfgang Amadeus Mozart's Piano Concerto No. 27 in B-flat Major, Pyotr Ilich Tchaikovsky's "Romeo and Juliet"). The diversity of ideas represented in the music can be explored by the CMM participants, identifying distinctive aspects of those ideas and then seeking and forming unifying concepts.

A supplemental listening session might also include listening to less familiar recorded music—classical (e.g., John Cage, Steve Reich); new age (e.g., Paul Winter, Kitaro); or world music (e.g., Indian, Indonesian, African, etc.). Popular forms of music (e.g., rock, metal, rap, Broadway, country western, etc.) may also be included as a basis for comparison/contrast.

The participants in a CMM listening session should try to find and share connections between musical ideas found in the differing musical genres. Participants might also form general concepts about the music heard or about the experience of listening, which they can then apply in their next playing experience.

11

CMM STEP 4—QUESTIONING

The questions that "engage our thought" on a daily basis reflect our life purpose and influence the quality of our lives.

—Michael J. Gelb, *How to Think Like Leonardo da Vinci*

After the performing/recording steps and/or after listening to the recorded playback, the fourth step of creative music making begins. Questioning in a CMM session means raising questions, but it also means thinking, discussing, assessing, contemplating, and evaluating those questions. The main challenge for participants is to know what questions to ask.

In CMM sessions it is recommended that questioning and discussion occur in a round-table format, with all participants sitting in a circle facing the center. A good technique is simply to go around the circle in order with everyone contributing to the flow of thoughts and perceptions. For each successive question, a new first respondent can be selected to begin the clockwise (or counterclockwise) order of follow-up responses.

The "questioning" step might begin with questions introduced at first by the facilitator. In a short time these questions can be raised and addressed directly by participants as they become internalized and self-directed, without any prodding from the facilitator.

QUESTIONS FOLLOWING THE CMM PLAYING STEP 1

Immediately following a performance in the playing step of CMM, questions of the following kind can be directed by the facilitator to the players before they listen to the playback of their performance:

- What were you thinking as you played?
- How did you know what to play?
- How did you relate to the other performer(s)?
- Who was leading/following? Where?
- Did you like what you played?

Fig. 11.1 Questioning in a CMM session means raising questions, but it also means thinking, discussing, assessing, contemplating, and evaluating.

- What else could you have done?
- If you could go back, what would you have changed?
- Did you try to play any musical ideas (melody, rhythm, pattern, etc.) that are based on music you have heard before?

The facilitator, upon hearing the response of one of the players, might turn to another of the players and ask, "Do you agree? Is that what you heard?"

Another question, "What did you hear—what melodies, rhythms, voices?" might draw attention to a player's objective listening experience. This question can be addressed by simply recalling those identifiable elements of the music that were noticed while playing. All of the basic elements of music—melody, harmony, rhythm, tonality, and the like—can be noticed. An absence of one or more of these elements is also significant enough to be addressed in discussion.

The facilitator can ask the players a subjective question, which by its nature requires a personal response: "What did the music say to you—what was its mood?" The responses will reveal as much about the player responding as about the music.

One of the greatest powers of music is the power to affect emotion and mood. When one player describes a mood or emotion, the questioning can proceed to identify exactly what qualities in the music evoked such a response: "Was it the fast tempo that made the music exciting?"

After the players have given their responses, the same questions can be posed to the nonperforming participants. Similarities and differences in perceptions and observations can become subjects for further questioning and discussion.

QUESTIONS AFTER LISTENING TO THE PLAYBACK OF A CMM PERFORMANCE

Dialogue does not require people to agree with each other. Instead, it encourages people to participate in a pool of shared meaning that leads to aligned action.

—Joseph Jaworski, *Synchronicity*

After listening to the playback of a performance, all of the above questions may be revisited again, and more questions may be raised, such as:

- What were you thinking as you listened?
- Did the playback sound different or the same to you compared to when you were performing? In what way(s)?
- How did this music relate to other music that you know?
- Did you hear anything in the music that seemed familiar?
- How did the performers relate to each other?
- What challenges were faced by the performer(s)?
- Who introduced new musical ideas, and what were the responses?
- What in the performance did you like/dislike?
- What would you say to an audience to help them to gain a greater appreciation of this music?

A fun way to explore the subjective listening experiences of the participants is to raise the questions, "What title would you give to this music? What was it about the music that made you think of this title?"

The desired effect in addressing any of the above questions is that in their next performance participants will be better equipped to listen more perceptively and respond more musically, even though there may still be some insecurity and uncertainty.

Any participant's response can always be further developed by asking more questions: "Why? What? How? Does everyone else agree? Who disagrees?"

It is important to remember, though it may already seem obvious, that there are no wrong answers to any of these questions. The purpose of the questions is not to elicit "correct" answers; it is to encourage participants to think for themselves and listen in a more attentive way.

QUESTIONS FOR CMM SUPPLEMENTAL LISTENING SESSIONS

Not only African, Balinese, and Indian music, but also Javanese, Korean, and many others are having a strong effect on Western musicians.

—Steve Reich, *Writings about Music*

After listening to selected recordings of music, whether familiar or unfamiliar, questions like the following might be considered:

- What is important in this music?
- What are the main musical ideas in this music?
- What do you think the composer/performer of this music wants to communicate?
- What makes this music distinctive—what is familiar/unfamiliar about it?
- In what ways does this music compare with other music you've heard in this session?
- What did you hear that was new to you?
- Was there anything you heard that you did not understand?
- How can you come to a better understanding of the music?
- How can a better understanding of this music help you in your playing?
- Can your intuition be helpful in coming to an understanding of this music?
- What was going on in your mind as you were listening? Did you listen in the same way you listen in other (musical or nonmusical) situations?
- How can listening to this music help you to broaden the range of possibilities in your playing?

UNDERSTANDING MUSIC

> *When we appreciate music through understanding, we do not take meaning from music. We give meaning to music.*
>
> —Edwin E. Gordon, *Rhythm*

Music is valued around the world for its ability to enrich people's lives. It can accomplish this by reminding listeners of their spiritual connections to other people, including people of other times and places.

Of course, music is not the only means by which people can connect. All of the arts and any other human activities that are practiced with an attitude of reverence can serve as vehicles through which people interact, experiment, reflect, and share in learning about themselves and others in a way that, unlike some other activities, causes no harm or injury to anyone else. In music, important connections between listeners and the world around them can be made through an understanding of the music and its meaning.

Understanding in music can happen in a number of ways. For some it's a logical process—like a puzzle; a listener simply puts all of the musical clues together and comes up with an understanding. For others it might be a process of gathering facts about the history, context, composer, or style in order to determine what the composer or performer really wanted to express, or what a particular stylistic "school" or "movement" was really trying to say. In hearing classical music many listeners form their understanding of the music based on the program notes that have been written specifically by the composer or by some other musical authority.

Another way of understanding music that is widely accepted around the world involves "representational" connection. This is a conceptual framework in which a certain musical phrase, chord, or melodic motif is intended by the composer or performer to arbitrarily represent a specific idea, person, or emotion. In this conceptual framework an understanding of what the music represents is only possible once the listener is made aware of the arbitrary linkage.

A more direct kind of representational understanding occurs when a composer or performer imitates the sounds found in nature or in the human environment—bird calls, machines, wind, thunder, car horns, ad infinitum. Such representations are usually understood readily, without the need for nonmusical information to make the connections clear to the listener.

More commonly, musical understanding is derived from words that are sung or spoken as part of the music. In fact, in popular song, folk music, opera, and church music, the words in the text carry virtually all of the ideas that are supposed to be understood, and the musical context serves primarily as a way to enhance the emotional impact of the words.

All of these examples are methods of understanding music that require listeners to grasp the ideas and concepts surrounding the music as well as to recognize objective aspects of the music itself. Another kind of nonrepresentational understanding is possible, and it can happen as the result of a fascination with sound.

Unfamiliar sounds—in music and in the environment—can provoke a sense of mystery and heighten the level of awareness in the listener. Could this be a holdover of the fight-or-flight instinct in which a strange or unfamiliar sound could signify a threat? In listening to music it is possible to take good advantage of this involuntary response by using the heightened state of awareness to listen more deeply and to perceive the complexities of the sound.

In CMM, some degree of understanding can be gained through objective listening as described earlier. But there are other means that can contribute to the process—experience in performing, thinking about the performance, and discussing the performance with others.

FINDING MEANING IN MUSIC

"Is there a meaning to music?" My answer to that would be, "Yes." And "Can you state in so many words what the meaning is?" My answer to that would be, "No."

—Aaron Copland, *What to Listen for in Music*

Finding meaning in music is not exactly the same thing as gaining understanding. Ultimately, the only one who can answer the question, "What does the music mean?" is the one who asks the question. It is a very different question than the common one asked by novice listeners, "What is the music supposed to mean?"

Using the word *meaning*, as presented above, the second question is unanswerable, because only the person asking can provide an answer, and that answer will be particular to the person asking. It is an incorrectly conceived version of the question, which should be, "What does the composer or performer want me to understand?" This question can be answered in an objective manner.

Musical meaning can only be determined by individual listeners as each one reflects inward—intuitively or rationally—and responds emotionally. When the listener connects his or her thoughts and emotions to similar thoughts and emotions experienced elsewhere—in music or in life—meaning begins to take shape.

By definition, life experiences are unique to each person, and so is meaning. Meaning is personal and it is therefore strongly related to the universal questions, "Who am I?" and "What is my place in connection with others?" It comes from understanding oneself, and it cannot be imposed by another person, even by a composer or performer.

It is in the nature of the creative music making process to offer an environment where participants can work with others in playful activities that force the questions that can reveal who they are to themselves and to others: "Why did I do that? What do my actions tell me about myself? What effect did my actions have on others? What can I do the next time to change the things I don't like?" Such questioning is fundamental to good musicianship, especially when performing in an ensemble.

12

CMM SESSIONS AT THE UNIVERSITY OF MISSOURI–KANSAS CITY CONSERVATORY OF MUSIC

What follows in the next few chapters are edited transcriptions of two creative music making sessions presented by the author in the afternoon and evening of February 6, 2003, for Dr. James Snell, assistant professor of percussion at the University of Missouri–Kansas City Conservatory of Music (UMKC), and his students in the percussion department. It is recommended that the musical examples on the enclosed CD be played at the times indicated while reading the text.

The intention in presenting these transcriptions is to convey a better sense of the CMM process to the reader through involvement, however indirect, in a real event as it actually occurred. In much of the following transcriptions, the words actually spoken have been edited or paraphrased in order to make the text more readable. At the same time, every effort has been made to preserve the ideas presented by each speaker.

Along with Dr. Snell, the student participants were percussionists Brett Baxter, Brenden Bennett, Will Braune, Roger Caliman, Tom Kernan, Bill Solomon, Liz Stephens, Nick Urbom, and Sam Wisman.

Because creative music making is intended to be useful for musicians on any instrument, not just percussion, an invitation was extended to other instrumentalists at UMKC. Joining in the sessions were John Hillan on clarinet, Hunter Long and John Thieben on saxophone, and Ryan Wurtz on guitar.

Mixed instrumentation is generally preferred in CMM, because it enables the participants to learn about instruments other than their own. It can help players to grow "outside the box"—beyond the technical limitations and the practical concerns of their own instruments. That being said, however, there is absolutely nothing wrong with having all players performing on instruments that are the same or similar.

The afternoon session was two hours long and it began with a summary description of CMM. In the transcribed texts that follow, the names of the participating students will be omitted; they will simply be referred to as Player 1 or 2, or Listener A or B, and so on. The author will be referred to as the Facilitator.

Facilitator: This workshop is called creative music making. I want to take a few minutes at the beginning to speak to you, and then we'll get to the playing part. By the end of this two-hour session all of you will have played in an improvisation. I want you to clear your minds of the things you normally think about when you're playing and focus on other things: What's going on? What do I hear? What do I do now? Before we start you should know that in these improvisations there will be two rules. Rule number 1 is that you can play whatever you want to play, and rule number 2 is that you should try to listen to what you are doing and to the music that's being made. You should try to connect to the music in some way. How does that happen? I don't know, but we'll find out. However, there's no penalty if you break a rule. If you don't listen, nobody's going to give you a traffic ticket.

[*At this point, participants were asked if they had ever performed an improvisation in a concert, and a few of them raised their hands. An assessment was made to determine what instruments other than percussion were going to participate in the sessions.*]

Fac.: We'll have mixed ensembles. Would one of the percussionists like to improvise with me?

[*After a show of hands a volunteer is selected.*]

Fac.: I'll take the vibraphone and you can take the marimba. Do you have sticks you can use? You can use one, two, four, or ten [*laughter from the percussionists*].

Fac.: We'll announce our names for the recording. You say your name; I'll say mine, and we'll just wait a few seconds until it becomes silent in the room, and then we'll see what happens. That's all there is to it.

13

TRACK 4—DUET IMPROVISATION

Play CD Track #4—Duet Improvisation (Time: 4:29)

Players: Bill Cahn, vibraphone; Nick Urbom, marimba.

Facilitator: Nice job! Before we listen to it in playback, think about what you were hearing. What did you notice? What did you think as you were playing?

Player 1: I was feeling a lot of longer, more legato things happening in the music, so I went along with that. Sometimes I'd really get into what was happening right in the moment, and sometimes I'd think like someone listening from the outside. I'd go in and out of that. I noticed that if I played longer, clustered sounds, as opposed to staccato random sounds, that I could stay more involved with the music.

Fac.: How much were you aware of what I was doing?

Player 1: I felt like I was aware enough, that I was letting [what you were doing] affect, but not dictate, what I would do.

Fac. [*to all participants*]*:* Does anybody disagree with that? Did anybody observe it differently?

Listener A: I heard the same thing—just the long sounds—especially the way you [Facilitator] started, and he [Player 1] played off of that with chords.

Fac. [*to Player 1*]*:* Did you feel like you were leading or following?

Player 1: I felt like we switched back and forth a couple of times.

Fac.: Let's listen to the playback.

[*All participants listen to the playback of CD Track #4*]

Fac. [*to Player 1*]*:* What did you hear in the playback?

Player 1: I heard you [Facilitator] even more.

Fac.: I was probably louder. [*Laughter.*]

Player 1: Beyond that, I heard you making more happen, as opposed to letting more happen.

Fac.: Does anyone disagree with that?

Listener B: You [facilitator] had harder mallets on the vibraphone and he [Player 1] had soft mallets on the low end of the marimba.

Listener C: Nice contrast!

Fac. [*to all participants*]*:* What would you call the piece—what title?

Listener D: I would call it "Two-and-a-half Pieces."

Fac.: Why?

Listener D: Well, there were two sections and a recap, I think.

Listener C: I found it interesting that dynamically, you were very much in tune with each other, whether you were listening to each other or not. Dynamically, there was this amoebalike shape. There were times when one of you momentarily emerged from the texture. Also, it was interesting how the piece ended. I'm always curious to learn—when there are two people not speaking, how do you know it's over? But, it was clearly over when one particular sound came in, and I think everyone here knew that the piece was ended. It was really interesting how the music just comes to that kind of conclusion. [*To Player 1*] Did it feel like it was over?

Player 1: Yes.

Fac. [*to Player 1*]*:* How did you know? What were you relying on?

Player 1: I was relying on when my gut told me it was time to quit. I didn't feel like playing anything else right there.

Fac.: Did anybody notice anything in the playback that was different from what you heard in the performance?

Listener E: I heard the vibraphone louder in playback than I did in real time.

Fac.: It could be the placement of the microphone.

Listener F: Because it's the second hearing, I know when it's going to end. I can spot the sections, so I can make more sense out of it.

Fac.: The two of us have never played together before, ever. Did anybody sense that this could have been a composed piece, or maybe a movement of a piece?

Listener G: Yes.

Fac.: Would it have been possible to tell that the music was composed or not, just by listening to it?

Listener H: No, I don't think so.

Listener J: There might be some clues, maybe.

Fac.: What would the clues be?

Listener K: It seemed like the first time I heard it, it just stopped. It was very silent. Then you [Facilitator] came back in. It could just have been a grand pause.

Fac.: Musically, we kind of said, "Hello?" carefully. "Hello! ... okay, now what?" [*Laughter.*] "Let's try this ... all right, we're talking ... now what?" That's what was going on musically, because we didn't yet know how we were going to make music together.

Listener L: Maybe in the past I would have thought that was improvised, but now that I've learned more about other composers—like [Karlheinz] Stockhausen and John Cage—it could have been in that style of composition. You really don't know if you're only listening to the playback whether it's really composed or not.

14

TRACK 5—TRIO IMPROVISATION A

Facilitator [*to all participants*]*:* Let's try another piece with a wind instrument and two more percussion players. You percussionists decide who will play which instrument. Which one of you wind players?

[*There is a show of hands and one is selected.*]

Fac.: I'll just review the rules: You can play whatever you want to play and you should try to listen, but there's no penalty for not listening. When you're ready, each player say your name. We'll have a moment of silence to clear the air, and then you can start the piece. [*To the Saxophone Player*] If you want to doodle a bit, just to get the instrument warmed up, that's fine.

[*The Saxophone Player warms up on his instrument.*]

Play CD Track #5—Trio Improvisation A (Time: 2:56)

Players: Will Braune, percussion; John Thieben, saxophone; Sam Wisman, percussion

Fac.[*to the players*]*:* What was going through your mind as you played?

Saxophone Player: It started off with no regular time. It started out with some mellow rolls, so I thought I'd begin with something on top of that. I was hearing harmonics. That's what I was hearing.

Percussion Player 1: I just began playing, hoping I could fit. I started listening to (musical) ideas, and to where people were going with them. If they went up I might go up with them, or I might go down. I tried listening to both other players at the same time. Sometimes I found myself just listening to one player, and as soon as I caught myself doing that or doing my own thing too long, I tried to open up my ears.

Fac.: Great!

Percussion Player 2: I remember thinking, "I've done some of this stuff before." The piece had a lot of clichés—which is cool! [*Laughter.*] It had a lot of rolls in the low register.

The saxophone took the lead, because it does have a "solo voice" compared to these two mallet percussion instruments. [*To the Saxophone Player*] I understand why you took the lead from the vibes.

Fac.: Let's do an audience poll. [*To the listeners*] If you thought you heard a cliché, raise your hand.

[*Some listeners raise their hands.*]

Fac.: So, it's not unanimous. [*To Percussion Player 2*] What you thought you heard didn't necessarily come across to all of the listeners.

Listener A: If I had listened without hearing the discussion we just had about the improv-or-composed question, I would have thought that the music was probably improvised.

Fac. [*to all participants*)*:* How do we feel about that? Raise your hand if you think it sounded improvised; raise your hands to commit to an answer.

[*Some participants raise their hands.*]

Listener B: I don't think you can tell.

Listener C: I agree.

Fac. [*to all participants*]*:* Let's listen to it again. [*To the performers*] Be thinking about what you hear in relation to what you heard as you were playing.

[*All participants listen to playback of CD Track #5*]

Percussion Player 1: In listening to the playback, I thought it was cool. For much of the time I couldn't tell the difference between the marimba and the vibraphone. They were just sort of going in and out of each other; the saxophone was really prominent. [*Laughter.*] John [on saxophone] has amazing ears and can just pick up notes, so I think he was not just playing random notes. A lot of what he played were actually notes out of the chords we were playing. I was following him all the way. On that playback I could hear it.

Fac. [*to the Saxophone Player*]*:* Were you doing that?

Saxophone Player: I was just listening to what the percussionists were doing most of the time and trying to fill out the sound.

Listener D: But it wasn't in the context of a harmonic structure, was it? It seemed more about each individual pitch.

Saxophone Player: Right!

Listener D: So your playing was very linear, as opposed to harmonic.

Saxophone Player: Yes.

Fac. [*to the Saxophone Player*]*:* Did you hear the cliché? [*Laughter.*]

Percussion Player 1: I think it started off that way, but then it wasn't.

Fac.: Have you three ever improvised together before?

Percussion Player 1: Not as a trio.

Fac.: Do you remember your first lesson on your instrument? Compare that with the way you can play now. This is your first improvisation together as an ensemble. Isn't it pretty amazing? [*To all participants*] Raise your hand if you heard any wrong notes. [*Laughter; no hands are raised.*]

Listener E: In the beginning I wanted to hear less from the saxophone trying to go for the overtones. [*Laughter.*] You definitely couldn't tell from the recording what he was after.

Fac.: Did it matter?

Listener E: No! It was cool.

Fac.: Listeners, how was it different now listening to the playback?

Listener F: Listening to it in playback, I tend to analyze it and to say, "This is A–B–A." Or, I'll make a concept like, "Does the beginning really match the end?" Is it good to analyze like that? Do we naturally gravitate toward analyzing the form, or is it good just to listen and not try to analyze?

Fac.: Good question! What's your answer?

Listener F: I think we naturally try to learn. In listening to that playback, I tried to put the music into a particular form.

Listener G: I don't know about forms, but I think what the trio played had a beginning, middle, and end, at least.

Fac.: Where do ideas about form come from?

Percussion Player 2: It's from all the music that we play that has a beginning, middle, and end.

Fac.: That would be my thought, too. We're all exposed to music that is structured in certain ways, so it's natural that even if we're not aware of any restrictions, we might simply do what we already know without even being aware of it because that's what we're exposed to. What this playing and listening experience can do for you is to help you expand what you have inside. It can enlarge your musical vocabulary. Try to listen to as many different kinds of music as possible and play as much as possible, the point being to recognize the limitations you have and to broaden your field of possibilities. [*To Listener F*] So, your question was a good one.

Listener F: When I listened to the improvisation as it was played, I was thinking about the middle, end, and beginning. When I listened to it on the recording, the music started and then it was over. It wasn't because I was thinking about something else other than form. I don't know how to explain it.

Fac.: It seemed different to you? [*To all participants*] If you agree with that, does anyone have an explanation?

Percussion Player 1: I disagree. When I was listening as we played live, I started to hear what was happening in the music and to ask myself, "Now where are we going?" When I listened to the recording, that's when I started to think about the form of the music, because I had already heard it once.

Listener D: I'm kind of interested in how the three of you worked physically, because if you listen to the improvisation on the recording, especially toward the end, there was a lot of interaction going on. But when you first played, it was as if there were "movement police" waiting outside the door, and if you moved at all, you were going to be arrested. [*To Percussion Player 1*] You moved to the top range of the instrument, but there was no sense of openness or comfort. [*To the Saxophone Player*] On the saxophone, you were like, "I'm not going to move." Less of that was noticeable on the recording than I thought. It sounded much more together and much more equally inspired than it looked when I saw it being played.

Listener H: I had a similar thought about the visual aspect of the music when I heard it during the playing. Maybe I wasn't listening all that much the first time because I was watching. I remember one time when I looked up at the vibraphone and the player seemed to be asking himself, "Which note should I go to next?" And at the end, when I saw one of the percussion players raise his hands to play a note and then lower them, it was as if he was saying to himself, "This is the end ... oh no, maybe I should play again. ... No, they're still going ... I guess I should play again." I realize now that I wasn't only listening to the improvisation; I was seeing it, too. For me, the improvisation really didn't go anywhere when I heard them play it. Then I heard it again on the recording, and I actually heard more when I was just using my ears.

Fac.: Very good point.

Listener J: When I was watching it being played, I paid more attention to the nuances of the body language than to the actual music.

Fac.: That's one difference between live performance and recorded music; it's the added visual dimension that makes a big difference. Here's one last question and then we'll go on. Close your eyes so your answer won't be influenced by what you see from the other people here. How many of you consider this improvisation—the one we just heard—to be interesting from beginning to end? Raise your hand.

[*Some participants raise their hands.*]

Fac.: How many of you found it to be interesting just for parts?

[*Some participants raise their hands.*]

Fac.: How many of you found it not to be very interesting at all?

[*No participants raise their hands.*]

Fac.: You can open your eyes now. It was about 50/50 for the first two questions. Nobody raised their hand for the third question. So you all found the music to be interesting in some way! [*Laughter.*]

Listener K: I have a question. For typical young children, what if the wind players use percussion instruments? I'm only asking because the percussion instruments seem so accessible. What if I were to say to a novice, "Here's a clarinet"? If I was trying to get a young kid to improvise, I wonder if that would work.

Fac.: With rank beginners and especially nonpercussionists, the problem is the nature of their instruments. With wind and string instruments it generally takes longer to get a good sound. With percussion instruments it's different. Nonpercussionists can just hold a mallet, walk up to an instrument like the marimba, and with a couple of practice strokes get a fairly decent sound. It's quite accessible. With very young kids, and even with non-musician adults, or with musicians who play other instruments, it's fine to have them play on percussion instruments. Another familiar instrument that is very accessible is the piano. In fact, on one of these improvisations maybe we'll have a percussionist play on a piano just to see what happens. You can also use [Carl] Orff instruments. Or, it's possible to devise homemade instruments or use found objects like tin cans, on which you can improvise fairly easily. So, there are ways of addressing the instrument issue with rank beginners. We're not facing that situation with this group.

15

TRACKS 6 AND 7—TRIO IMPROVISATIONS B AND C

Facilitator: Let's have two wind players and one percussion player this time. In fact, all of the wind players might want to take this opportunity to get your instruments unpacked and ready to play. Feel free to warm up for a minute or two.

Play CD Track #6—Trio Improvisation B (Time: 2:27)

Players: Roger Caliman, marimba; Hunter Long, saxophone; John Hillan, clarinet

Fac.: Great job! Before we listen to the playback, do another quick improvisation and make it different in some way. Use your imagination. Make it as different as possible from what you just played, whatever that means to you.

Play CD Track #7—Trio Improvisation C (Time: 1:49)

Players: Roger Caliman, marimba; Hunter Long, saxophone; John Hillan, clarinet

Fac.: What was going on? What were you aware of?

Player 1: Well, in the first improvisation especially, I was really trying to think about what I was doing, and sometimes I was just feeling.

Player 2: I was listening really hard to both of the other players, and I was trying not just to react to what they were doing, as much as to introduce some new ideas. I didn't want to just wait for them to do something and then imitate it.

Player 3: In the first improvisation I just tried to listen better. In the second improvisation I relied more on watching what the other players were doing, especially on the marimba, and then tried to match that in the style.

Fac.: Do any listeners want to comment on what you observed or heard?

[*Many hands are raised.*]

Fac.: Whoa … right! [*Laughter.*]

Listener A: One thing I noticed was that at first the two wind players were breathing together, which was actually interesting for quite a while.

Fac. [*to the wind players*]*:* Were you both aware of that?

[*There is a mixed response from the two wind players.*]

Listener B: In the beginning of the first improvisation the players were unsure about what was going on. By the end of it they were cooking; there was some cool stuff. The chords in the marimba probably sounded the most composed.

Listener C: It's interesting that the musical ideas seemed shorter. One player would play something and immediately someone else would change and go along with it.

Listener D: Do you think that's related to the players listening to each other a little bit more, instead of doing their own thing? If so, I'd have to agree. These improvisations were much more segmented. One would play a musical idea and another would pick it up. Then another player would imitate or play off of that, but immediately look for the next idea to pick up.

Listener E: I thought the first improvisation was really cool. The marimba was playing chordal ideas, and I assumed both wind players would play countermelodic lines, but instead they'd find an interval or something, and then play off of it—basically playing chord fragments. I wasn't expecting that, and I liked it. It was as if the two wind players were functioning as one chord, and the marimba was functioning as another one.

Fac. [*to the wind players*]*:* Were you both aware of that?

Player 2: Not me.

Listener E: I mean, if you had each been working independently on a separate audio track, the music might not have seemed chordal. Maybe you were each just thinking of slow melodies, but it seemed chordal.

Listener F: It seemed to me that the marimba wasn't instigating the two wind players. The music was more about conversation. Two players started, and the other answered with a short snippet.

Fac. [*to the wind players*]*:* Were you aware of that?

Player 3: I remember when the marimba and I were playing a rhythmic motif, and the other player did a longer melodic line underneath.

Listener G [*to Facilitator*]*:* As you were saying, it'll be interesting to hear the playback. But when the facilitator told you to do something different, it started off very differently. Then you'd go right back to where you were before musically. [*Laughter.*]

Fac.: It was like you were each thinking to yourself, "I'm going to dive into the pool … [*Facilitator makes a hesitant diving motion and then backs off*] … whoa, not now!" [*Laughter.*]

Listener H: In the second improvisation you could see that there was a lot more eye contact. It was interesting to watch how that affected what was happening with the music. That's why I associated it with sounding very segmented. One player would play something and then look up to see what the reaction of the other players was. It was very interesting.

Fac.: Let's listen to the playback. Thank you players. [*Applause.*]

[*All participants listen to playback of CD Tracks #6 and #7*]

Fac.: Does somebody want to suggest a title for that second improvisation?

Listener A: "Getting Tired." [*Laughter.*]

Fac.: Okay, any other?

Listener B: "Make Up Your Mind." [*Laughter.*]

Fac.: How about, "Search for the Lost Note"? [*To all participants*] What did you notice listening to the playback?

Listener C: I really liked what the clarinet was playing. I heard a little musical greeting that was really neat, and I also liked the part where both winds played the same note.

Fac.: And, there was another spot where all of you came in with that little trill. That kind of moment in an improvisation I call a "consonance." You didn't plan to play those notes in unison, but somehow, when you're really listening hard, things like that happen; things come together for a moment. It happened in all the improvisations, but this is a clear example of consonance—both coming to the same note at the same time, and noticing, "Whoa, we're together here." It's like a shock, and then the players naturally try to cling to it. With more experience in this improvisation process—remember, this is only your first time improvising together—the more those consonances happen. They can't necessarily be explained, but they do happen. Ultimately, the goal with this kind of music making is to get *there.* Call it "in the zone," as the sports people do, or use whatever expression you want to use. You want to get there, and then you want to stay there—to be deeply engaged in the music and fully involved with the other musicians in prolonging this consonance, whatever it is. It's good that you heard it in the music because I wasn't sure that it was noticed.

Listener D: One of the interesting things about the last trio was when the facilitator said, "make it as different as possible," it became disjunct. It was interesting that the first thing the players thought of as different was the length of the phrase, not dynamics or anything else.

Listener E: That's what I thought during the second improvisation, too. When I was hearing the playing live, I had my own expectations about what would be different. But when I listened to it in playback I thought, "This is very different, just in a way that I was not expecting."

Fac.: Expectations are good things to talk about, because one of the great things you can do with expectations is work with them as musical tools. You can thwart expectations in an interesting way, or you can meet expectations in a certain way. It all depends on your inclination at any particular moment in the music. It's not necessarily a decision you think about, but it is part of the process here. [*to all participants*] Those expectations come from where?

Listener F: From your own experience.

16

TRACKS 8 AND 9—TRIO IMPROVISATIONS D AND E

Facilitator: Who hasn't played yet? We need the guitar for sure. Are there any wind players who haven't played yet? Let's go with three, then. Say your names, take a moment of silence; then begin.

Play CD Track #8—Trio Improvisation D (Time: 3:49)

Players: Brenden Bennett, marimba; Liz Stephens, vibraphone; Ryan Wurtz, guitar

Fac.: And now, play another improvisation—something completely different.

Play CD Track #9—Trio Improvisation E (Time: 2:53)

Players: Brenden Bennett, marimba; Liz Stephens, vibraphone; Ryan Wurtz, guitar

Fac. [*to players*]*:* Where were the musical ideas coming from? How many musical ideas were there in the second piece?

Percussion Player 1: I was responding to what the vibraphone was doing. If she was playing on the sharps and flats [black keys], I was trying to play down on the naturals [white keys].

Guitar Player: I thought everybody was trying not to imitate each other in the second improvisation—trying specifically to do something different from the first piece.

Percussion Player 2: It took me a while to really sync with the marimba.

Listener A: I see a pattern developing when we're using the technique of "… and now for something totally different." The first improvisation always seems to be a little bit more lyrical, with more complete, full thoughts from the players. The players are listening to each other at a deeper level and the music is not quite so fragmented. When the facilitator says, "Now for something totally different," the players all of a sudden look up at each other, and the musical ideas become shorter and fragmented. *Different* seems to mean *quicker* almost instantly. The players change to a new mode, but then they ease back into the way they were playing before. Almost every second improvisation was done this way. They have turned out to be in that pattern.

Percussion Player 1: That observation described exactly what I did. The first improvisation was slow because it's an easier way to play. Long tones provide you with the time to work with them either quickly or slowly. When the music is short and fragmented, if you're really quick on your toes, then you get the ideas bouncing back and forth more rapidly between each other. So, playing the long tones in the first improvisation is just an easy way to understand what each of the other players is doing.

Listener B: That's exactly how I felt earlier, when I was up there playing.

Listener C: Here's a different idea: I felt like I wanted the players to try to develop their melodic figures more. I don't think the music really took off anywhere. The guitar seemed to be in charge, a lot of the time, melodically. The marimba didn't seem to have any idea about what he was playing; In the second improvisation the vibraphone seemed to be leading. The guitar was trying to inject musical ideas, too. The vibraphone was leading and trying to bring the others along. The minor seconds were really harsh tones. I wanted the players to expand on the minor seconds, maybe to an interval like a fifth, or whatever. Again, I could never really hear what the marimba was doing. Maybe its more audible on the recording.

Percussion Player 2: I find your comments interesting that the guitar was leading the first improvisation, and the vibraphone was leading in the second improvisation. The guitar started the first one and the vibraphone started the second one.

Listener C: That's exactly what I heard.

Listener D: I noticed that having a rhythmic pulse seemed to be dangerous to the players. Maybe they avoided it because there was a risk that someone would lose the pulse and mess up the music. Especially as percussionists, we're so used to having a steady pulse. In lots of music I'd just like to hear a pulse. If it's lost, so what! It's interesting that no one has played a pulse.

Fac.: I'm wondering if the very first improvisation has affected those that have followed. It probably has. Let me present some other possibilities here. The players do not need to have the facilitator say, "Now play something completely different." All of the players have that option at any time. What I'm hearing is that everyone's trying to be in harmony and in sync with everybody else. It is entirely possible for any player to say, "No! I'm going in an entirely new direction," and bring new ideas into the music. That's one way of keeping the music interesting. When someone offers a new idea in the music, then the issues for the other players are, will you go along with this idea, or will you cling to your existing idea, or will you inject yet another idea? That kind of interplay is just starting to happen in your improvisations, but not with drastic changes. You always have the option of saying, "This is boring. I'm going to do something different." Then the other players have to either buy into it or not. Or, you may want to rethink your new idea: "This is not working. I'm going to back off here a little bit." Or you may say to yourself, "I like how this idea is working. I'm going to stay with it for now and try to interest the other players in joining along with it."

Listener E: In both of the last two improvisations, it sounded like that kind of process was going on the whole time. Everyone was trying to lead but nobody wanted to follow, so they didn't. The ideas would just sit in place and no one would go along with them. Well, maybe they would for just a moment—a few seconds—but not any longer. Nothing was ever outside of the overall texture. It was as if they were trying not to play anything inappropriate. There was nothing in the music that sounded like it didn't belong there, but it sounded like ideas were coming from all directions. I was thinking, "Should I listen to *this* idea ... or should I listen to *that* one." However, it didn't really matter whether or not the ideas settled into anything. It was just interesting to me that everyone was keeping their distance musically from the other's ideas.

Listener F: I noticed that the guitar was working the volume pedal, in the second improvisation especially. It was as if he was trying to get the other players to listen to his idea. [*There are nods of agreement from other participants.*] He would play some complex chord and turn the volume up and down.

Fac.: So, if other people don't buy into your idea, one choice you have is to play it stronger again. Any player can be insistent about an idea or back away from it; those are options all players have at any time. And there's no obligation to be harmonious—to even have a connection with the others. There's no obligation to have any perceivable connection. You are totally free at any time to take the music in another direction if you want to, and then to see what happens. A player does not need to be in any particular musical relationship with the other players. It's nice when it's consonant, but it can also be nice when it's dissonant. It's interesting when you're listening to someone's musical idea; but it can also be interesting when you're not listening. Do any of you have thoughts on this subject?

Listener G: I had a friend in high school who used to play the drum set. On one occasion we were recording some music, and he played a time groove that everyone thought was really cool at first. But then we recorded it again a few times and it was always exactly the same thing. We found ourselves in a rut, so we became discouraged and stopped recording at all.

Fac.: The problem is how to get out of that rut. Even when you're playing within a piece, you can get into a rut—an inability to break away from a musical idea. In fact, within a musical phrase you can get into a rut. The thing is, when you notice it—when you say, "The music isn't happening for me now"—then you have a choice: change, or connect in some way to make it happen now. That kind of "being in the music" is the real issue: "What do I do to keep that center, that focus, that dynamic of music making, whatever it is?" A player, at any moment in these improvisations, has the option of changing what he or she is playing or connecting in some other way. Listening is not only about what's going on in the big picture, or what's going on inside you, but also whether you believe in this music. Do you find it interesting? Is it boring? Does it not feel right in some way? So, "What can I do now? Where do I go?" Those are the kinds of issues that a player faces. This workshop is only the first baby step. With a little bit of experience, this way of making music can be absolutely as good as music making gets. And, you all sound fantastic—a really good start! Let's listen to the playback.

[*All participants listen to playback of CD Track #8*]

Fac.: That had a wonderful mysterious quality to it all the way through. What a great combination of instruments. What thoughts went through your minds as you were listening to that improvisation?

Percussion Player 2: We were gravitating toward a tonic pitch, which is a natural thing to do. That's what I was aware of the whole time. I heard a repeated melody on top of what the guitar was playing. It gravitated toward the tonic; it always returned to that.

Guitar Player: I agree, because I decided to start directly with a chord rather than another sound. Wherever the other lines were going, I would listen to it fully and return back to that chord.

Percussion Player 1: I was trying to match the chord as we played.

Fac. [*to Percussion Player 1*]*:* Were you trying to match notes or were you trying to match feelings, or what?

Percussion Player 1: I was trying to stay inside the chords—to play notes that sounded like they belong in the chord.

Fac. [*to Percussion Player 1*]*:* Were you fully aware of the chords?

Percussion Player 1: Yes; I was trying to match the notes, but sometimes it came out otherwise.

Fac.: That improvisation would be a fantastic first track on a CD. In fact, making a CD could be a future project to follow up on this workshop. Let's take a moment to envision where this experience in improvisation can lead you. It would be entirely possible for you to play an improvised piece in the context of a recital. It would be a perfect place for it. It can be quite a thrill to make up music on the spot in front of an audience, especially in front of a paying audience. [*Laughter.*] There's a certain amount of focus and a certain amount of "ear factor" in an improvised live performance that heightens the musical experience. I would encourage you to try it. You needn't worry about failure, because there are no wrong notes here. No one has played a wrong note here today. The issue is to make music and play something that's interesting to listen to. Another thing that can result from this creative music making process is the creation of a CD. In NEXUS we've made CDs, not only with our percussion group alone, but also with other instrumentalists—with clarinetist Richard Stoltzman, with bassist/stick player Tony Levin, and with Eric Robertson, an organist. You can create an entire CD of just music you've improvised. You can have fun choosing titles for the pieces. Then, you can give the CDs to your friends for Christmas. [*Laughter.*] It could be a fun project to do. Develop a body of improvised pieces recorded for a CD, select the best ones, and put them in order. Let's listen to the second improvisation.

[*All participants listen to playback of CD Track #9*]

Fac.: What do you think about this improvisation?

Percussion Player 2: I think I was leading it because I started it. I listened more to the guitar at first, and then I started to play off of whatever he was playing.

Guitar Player: When we first started this improvisation I was trying to play something completely different from the first piece. I asked myself, "What could be the most different thing we could do? Should we start playing some dissonant notes?" At first I didn't know where the music was going. I wasn't sure what to do with the musical ideas. In listening to the playback, it sounded more together. Toward the middle of the improvisation we started connecting on some ideas. But when we were playing live, I remember just thinking, "What can I do with this?"

Percussion Player 1: I didn't really connect with the other players' ideas until after I came in. I was just randomly striking the instrument. I wanted the other players to match what I was playing, but I just didn't fit in.

Fac.: Would anyone else here raise your hand if you think he (Percussion Player 1) didn't fit in?

Percussion Player 2: Maybe the reason you feel that way has to do with the instrument choices. The guitar has a wash of sound, and the vibraphone has a sound that rings a lot. The marimba's crisp and short, unless the player is rolling, especially in the lower register. The marimba isn't making the same fullness of sound. Maybe that's why you felt the way you did. It seems like you feel bad for not fitting in, but I don't think you have to. The guitar and the vibraphone don't have to be playing the exact same notes. Sometimes we play around the same few notes, but even that doesn't have to happen. I wouldn't feel bad if you're not playing exactly what another person's playing.

Fac. [to the listeners]: Are there any other thoughts?

Listener A: I really liked the first improvisation. During the second one, as I was watching it being played, I was thinking, "Where is this going? What's the point?" Now, when I'm listening to it in playback, I can actually visualize movie scenes; this piece could be a soundtrack. The marimba player was saying that he didn't know how he was fitting in, but in listening to it on the playback, I wouldn't necessarily want to hear it any other way. I thought the randomness of everything worked really well. In the end it came to a quasi-resolution, but a kind of resolution in which nothing is found. But a resolution was there. I liked it.

Listener B: I didn't like the second improvisation. I don't know why. It began with a new pattern of loud, short phrases, and then it just fell back into the pattern of the earlier improvisations, but it was less inspired.

Listener C: I definitely thought that the first improvisation had more direction to it. I agree with you; I liked the first one better.

Listener D: Upon listening to the playback, I heard a little more of the players' personalities in the second improvisation. When watching the players live I didn't see that. Upon listening to it, it seemed like the players were being themselves more.

Fac.: I have two comments to make, and then you respond. First, the players are not obligated to play anything. If a player thinks, "There's nothing touching me here," then it's not necessary to play anything at all. That's an option for any player. On the other hand, there's nothing wrong with continuing to play when you're not feeling 100 percent involved in the music. A player always has the option of not playing. Second, a player can take a contrary approach to playing. This is one way for a player to expand a musical idea and to get outside of a rut. For example, a player might think about playing a major chord. What the player can actually decide to do is to be contrary—to play a major chord, but alter one or two of the notes. When a player has an impulse to do something, the player can decide to do exactly the opposite. The interesting thing about this approach is that the player can learn things that weren't known before. If you feel inclined to play loud, you can play soft instead. If you feel inclined to play soft, you can play something grossly loud. That's a method of injecting something entirely different, even as you're playing. It's a way to inject new possibilities into an improvisation.

17

TRACKS 10 AND 11—TRIO IMPROVISATIONS F AND G

Facilitator: Who hasn't played yet? Maybe there can be two players on a marimba, or if you see any other instruments in this room you want to play, go ahead. I'd like to hear a piano in one of these improvisations.

Listener G: I can play piano.

Play CD Track #10—Trio Improvisation F (Time: 3:46)

Players: Brett Baxter, percussion; Tom Kernan, percussion; Bill Solomon, piano

Fac.: So far, the pieces all seem to be about the same length of time. The pieces can be longer or shorter, whatever. They don't have to be exactly three-and-a-half or four minutes. What I want to avoid is limiting the possibilities for these improvisations. We've done that already in certain ways, including the length of the pieces. Let the music go where it goes. If the other players stop and you still want to continue playing, you're totally free to do so. On the other hand, if you think the music is over, you can stop and let somebody else take it wherever it goes.

Play CD Track #11—Trio Improvisation G (Time: 2:37)

Players: Brett Baxter, percussion; Tom Kernan, percussion; Bill Solomon, piano

Listener A: It seemed like the piano functioned as an accompanist. Maybe that's a natural role for the piano. It really sounded like a composed piece that could have been written. One percussionist would play and then there would be a little piano response. Particularly at the beginning, one percussionist played off of the idea that the other percussionist presented involving quick runs in the high register, which was good.

Fac. [*to the percussion players*]*:* Were you aware of that?

Percussion Player 1: Yes, actually. I didn't think of the piano as an accompanist. I think the relative positions of our instruments and how well we could see each other were important. In much of the beginning I was responding to ideas I heard from the other percussionist. So, "Damn the accompanist!" [*Laughter.*]

Listener B: For a while there was just a pulsing phrase. One percussionist was doing little rhythmic figures to keep the pulse going. The piano played a melodic line against the pulse. Then everyone moved away from the pulse and came to a close. It was a cool improvisation.

Listener C: In the first improvisation it seemed like both percussionists started off with wild ideas, alternating back and forth. I don't know why this came to mind, but it was like two parents having a conversation with an obnoxious child. [*Laughter.*]

Listener D: There was a nice change in the beginning of the first improvisation. Everybody was having a good time in the beginning, but then everything went serious all of a sudden. The players' facial expressions changed from the way they were at the beginning. With a lot of the improvisations we've heard, the players try to change the existing mood, but they keep coming back to it.

Fac.: When we play again, I would like to hear a little more risk taking. That's missing right now—the injection of something completely different, or something that you're not sure is right. We don't know what "right" is yet, but I sense that we're still reluctant to be "wrong." All we're really trying to do here is to have fun. Let's not make what we're doing too serious. Now, we can hear the second improvisation.

[*All participants listen to playback of CD Track #10*]

Percussion Player 1: I liked the improvisation more when we were playing it.

Listener A: I liked it when we were listening to the playback. [*Laughter.*]

Listener B: It started with kind of an introduction, and then it changed as if the piece was really supposed to start at that point. Maybe it's just me trying to organize the music into some kind of form. Actually, I did hear these things when it was actually being played, not just when I heard the playback.

Percussion Player 1: There are things I heard on the playback that I don't remember hearing at all when we were playing. There was a lot of stuff happening.

Percussion Player 2: It's hard with three different players, when there are that many different ideas coming so quickly, to do something that is good.

Listener C: Especially, when the second improvisation began, one player started and then the other ideas all happened at once.

Percussion Player 1: It was like several people talking all at once. If that had happened in a conversation—three people starting to talk at once—it would have been chaotic. In this music it became something that the players all shared.

Fac.: In what ways can a musical idea be developed? How can an idea be preserved? Earlier we talked about being insistent. What can be done? Are there any ideas about that? [*To the Piano Player*] Play a musical idea.

[*Piano Player plays three ascending notes on the piano.*]

Fac.: Okay, what can you do with that?

[*Piano Player plays three different ascending notes, then three descending notes, then three more ascending notes.*]

Fac.: What are you doing?

Piano Player: I'm using the same idea—three notes.

Fac.: What else can you do?

[*Piano Player plays three ascending chords.*]

Fac.: What else can you do?

[*Piano Player plays three broken ascending chords and single notes.*]

Fac.: What else can you do?

[*Piano Player plays three arpeggiated chords and single notes.*]

Fac.: What else can you do?

[*Piano Player plays two single notes.*]

Fac.: You can also repeat exactly what you've played. Or, at least you can try to repeat exactly what you played. Sometimes you can't remember exactly what you did. You might try to repeat an idea, but come up with some kind of variation. So, play the three notes again.

[*Piano Player plays three ascending notes on the piano, and repeats them several times.*]

Fac.: Vary it in some way.

[*Piano Player plays three notes on the piano, and repeats them several times in differing order of sequence.*]

Fac.: So, you can work with your ideas more. What's happening up to this point—and it's typical of these first improvisations—is that the players are not taking their ideas and working more with them. It is possible to bring the other players on board with your ideas, or to further refine your ideas yourself. What's happening now is the ideas are being shot out. In the case of the rhythmic ideas, they are ideas you can hang onto and agree upon. The thematic or tonal ideas worked very well in the guitar piece, where there was really not a lot of activity going on. The first improvisation was very static, but there was a lot of hard listening and a willingness to participate in what was heard rather than to change it in some way. However, the players can also take their ideas and work with them somehow to create more interest in the music—not only for the listeners, but for themselves as performers.

Listener D: If you hear something you like, and you're trying to attract other players into your idea, you can first be aware of everything that's happening and then try to connect with what another person is playing. You can connect with what they're playing for a second and then return to what you were playing, so that it draws their attention to your idea. It's like you're walking along with them, but actually, you're coming back to your idea. I think that tactic would be noticed.

Percussion Player 1: There's also an idea that I had early on and tried. No one seemed to notice it. Then later on in the improvisation people started to hear it. Maybe if you minimize things—play less and play softer—people are challenged to search a little more.

Fac.: Congratulations to all of the players. I hope that when you have an opportunity to listen to these recordings in the future you'll be pleasantly surprised.

18

TRACK 12—EXPANDING QUARTET IMPROVISATION

Facilitator: What we're going to do first in this evening session is an expanding quartet improvisation. The first group of players will consist of four people. On an arbitrary hand cue, which I will give, we'll add a second group of four, and on another arbitrary hand signal we'll add the third group of three players.

Play CD Track #12—Expanding Quartet Improvisation (Time: 16:14)

Players: Group 1—four percussion players start; Group 2—four percussion players join in; Group 3—three players (saxophone, two percussion) join in.

Fac.: Thanks for that ending stroke on the wind chimes. [*Laughter.*]

Listener A: I thought everyone did a good job. Listening to those long tones was fun.

Listener B: It was pretty intense.

Player 1: I had mixed feelings about the number of people playing. At one point I liked having so many players, because there were so many things to listen to, but at the same time there were places where I was struggling to hear everything, especially at the end when everyone was playing. I could see the player on the almglocken [Swiss bells], but I couldn't quite hear what was being played. There was one point where I saw the clave [Cuban sticks] player, and I was trying to connect with it. I thought we were together for a while. I didn't know if he even realized it, because there was just so much going on. It was great to have all those options, but at the same time I was lost a little toward the end.

Listener C: I really enjoyed when the chains were dragged on the floor. It made me think of sounds at the beach. The group was playing all of those sounds, and I just envisioned myself sitting on a California beach. Then, I really enjoyed the rhythmic groove. There were great ideas on the vibraphone. It sounded like a television soundtrack. [*Laughter.*]

Fac.: Nice piece; well done! You were all listening as you played. Why don't we listen to the playback?

[*All participants listen to playback of CD Track #12*]

Fac.: That was really enjoyable.

Listeners [*in various forms of agreement*]*:* Yes … it was good.

Listener A: I think I'd come here to hear that! [*Laughter.*]

Listener B: It was so much fun!

Fac.: At one time, the late 1960s or early 1970s, there was an ensemble of percussionists called M'Boom that played this kind of improvisation. They were based in New York City and Max Roach was the director. I can easily imagine this last improvisation as being attractive to an audience at a Village coffee house. I might even pay an admission charge to hear it. [*Laughter.*]

Listener C: Earlier in the sessions we talked about the gravity-force nature of a rhythmic groove, and how it so readily grabs listeners' ears, and immediately involves them in the music. It can have such a great potential for the development of a musical idea. But, I also understand the observation earlier that it's difficult to get out of a groove. Once it's established, what do you do? My impression is that up until now, everyone's been trying to avoid the dreaded groove. But, I think this is an area we can exploit. [*To Facilitator*] Please talk a little bit about your own experience—just the nature of a groove, and maybe some ways out.

Fac.: I don't think I can say anything you don't already know. Let me ask the people here, "How do you get out of a rhythmic groove?"

Listener D: It's weird for me to think about that question, because I'm always trying so hard to get into one. [*Laughter.*] As percussionists, it's really our job in most situations, especially on drum set or hand drums. It takes a real jolt to shock you out of it.

Fac.: Well, how would you do that? What kind of shock treatment would you apply? [*Laughter.*]

Player 2: I think there are at least two ways. There's the shock way of getting out, which is just stopping. The other way is fading out. As I was playing the almglocken, I could have faded out while playing a couple of notes or changing the rhythm, extending it just a little more and then gradually ending it.

Player 3: Our eighth-notes in the pulse were about the same speed, and our sixteenth-notes were about the same speed. To get out of the steady pulse we could have tried to make the sixteenth-notes faster or the eighth-notes slower to even them out.

Listener C: It's just that there's such a strong force in the groove! When the players got to the second groove in the piece and the drums started to come in, there were these snippets of rhythm where two of the players—even if they didn't mean to—would just fall into a groove. It was like they were swirling around the outside of a funnel, every once in a while getting sucked into the middle and then back out again. The groove was there; it was ready to deepen, and then it just dissipated as quickly as it appeared. But, it didn't dissipate naturally. It was as if the players were saying to themselves, "Oops! Go away, groove."

Player 3: I think what he's talking about is not a tornado funnel, but just a small whirlwind, where it gathers up dust around it. It seems to be growing, but in a second it's gone.

Listener E: I'm not sure if this can really get a player out of a groove, but it's something I used to do, especially with dance classes. The class that I worked with today went for twenty-five minutes at a time in the same tempo without a stop. But, usually I can switch the way I'm subdividing the pulses. If I'm playing eighth-notes for a while, I can subdivide in triplets and then just vary it. Sometimes it works and sometimes it doesn't.

Player 3: I agree with that. It's what I was trying to say but I didn't articulate it like that.

Fac. [*to Listener E*]*:* Do you mean that you play triplets or hemiolas?

Listener E: Yes, that's what I mean—hemiolas.

Fac.: In other words, you give weight to every third pulse instead of every second one.

Listener E: Yes, the tempo of the pulses stays the same. When I'd change to a new groove, the number of weak pulses changed between each strong pulse. That's all.

Fac.: Did everyone hear that two different observations have been made? One comment was that a rhythmic groove can draw you in with its power, and the other comment was that a groove can be limiting and make you want to get away from it. So was it A or was it B in this last improvisation? Or, was it both?

Listener F: I think it was both. It just depends on your perspective.

Fac.: The problem with wanting the groove to end is that you can make it end for yourself, but you can't get into the other players' minds to make them stop.

Listener G: It's like a mob mentality! [*Laughter.*]

Fac.: Well, there's certainly some kind of connective mentality present in a groove—or maybe a lack of mentality. [*Laughter.*] I wouldn't necessarily say "mob mentality" but there is some connection between players going on, and the nature of that connection is really what music is all about. It's what we're exploring in this workshop: What's going on between people? Who's asserting and who's backing off? Who's out in front and who's stepping on other people's toes? Who's afraid to get in? Who's doing what? From the perspective of these kinds of questions, this was a very interesting improvisation.

Player 4: I think the formal switch between different groups was good, because it was an infusion of energy more than it was a change. There were different instruments and it was a change. But, there was also a different kind of energy in each section of the improvisation. It wasn't really noticeable until after the switch. Even though all of us had our eyes open and we saw the large cues, there was no sense of stopping and starting. What impressed me was that for fifteen minutes the energy level was unbelievably strong. We've never done this kind of playing before, and after fifteen minutes, you'd think the energy level would be waning. But, this improvisation didn't have any waning effect. In fact, it was really interesting. Toward the end I stopped playing for a little while. I was done playing what I needed to play before everyone else was done, so I backed off and finished. It was also amusing listening to the piece come to an end. It was as if there were several players each trying to be the last to play a sound. [*Laughter.*] There was a moment before the end where I thought, "This is it! This is where it's going to end!" and then there was more playing.

Fac.: Who was going to have the last word? [*Laughter.*]

Listener C [to Facilitator]: Maybe you've had more experience. Does the ending, after the players have had more experience playing together, usually become more apparent?

Fac.: No, it's different every time. But, I can say that the most rare kind of ending I've experienced is where everyone is playing loudly and actively, and the music ends in abrupt synchrony. That usually doesn't happen. It has happened on occasion, but when it did, everybody looked at each other with surprise—"Whoa! We really ended together! Unbelievable!" [*Laughter.*] I can't account for why that happens; it's very rare. Usually, there's a winding down of activity, even though it may be brief. That's what was going on here in this last improvisation. Eventually though, each player says, "I'm out of here!" [*Laughter.*]

Listener H: It was very interesting listening to this last improvisation and thinking back on all of the improvisations we've played today. It seems like they've changed ten million times over. Remembering the first improvisations and comparing them with this last one, all of a sudden everything seems so much more appealing to the ear. It's possible to hear more and to understand why particular sounds were noticed or not noticed, or why a particular sound was played or not. I think it's just an incredible change and it can really be heard in the last playback.

Fac.: Quite a leap in a few hours!

Player 5: Sometimes, I'd hear somebody just playing a sound and letting it ring, or on the marimba playing maybe playing in octaves, just one sound. It was absolutely beautiful—just one note. There was a lot of other playing going on, but I'm listening to just this one sound.

Listener I: One of the things that impresses me the most, after listening back to the last improvisation, is how much the harmony doesn't matter; it just doesn't matter! I listen to it and I never think to myself, "Those sounds don't belong together." They *absolutely* belong together! When you hear them and you listen to them, they absolutely belong together. And, you know the music doesn't appear in a written chart anywhere. That was really nice to know, too, especially when the wind players and the other instruments performed along with us. I was unsure of how that was going to come together, but not at any time in the entire day, thinking back on it, did I think to myself, "Those sounds don't work well together."

Fac.: It's very healthy to have the experience of making music without thinking about all the normal stuff—"you grip the stick this way," or "breathe from the diaphragm," or "strike this part of the marimba bar." All of those kinds of things are important, and I don't want to say or imply that this experience in improvisation replaces any of that. It doesn't; it simply adds to it—it balances it. The kind of experience with improvisation you've each had today is not necessarily the answer to making you a satisfied musician; but, I do think it's a part of the answer. So I would encourage every one of you to continue with this process of creative music making. Remember, this is only the first day! The more you continue with it, the higher you can fly. The last improvisation was worthy of a concert performance, and I would like to encourage you to do in a concert what you have done in this clinic. I would also encourage you to invite players on other instruments to participate, so that it's not only limited to percussion. I wish all of you the best of luck, and I thank Dr. Snell and the University of Missouri–Kansas City for giving me the opportunity to be here with you today.

Note: In the months following the UMKC workshop, a number of follow-up sessions took place under Dr. Snell's guidance. He reported that he saw very tangible results in his students. Comments from the percussion students themselves indicated that they felt more comfortable with their instruments when playing and that they were listening more actively in their ensembles.

19

CMM IN GRADE SCHOOL CLASSROOMS

We have to establish already in schoolchildren the belief that music belongs to everyone and is, with a little effort, available to everyone.

—Zoltan Kodaly, quoted in Ian Crofton and Donald Fraser,
A Dictionary of Musical Quotations

Creative music making as a pedagogy was developed primarily for musicians, but for many of the same reasons given above, it can be of value in other applications—in a grade school classroom or an adult continuing-education class, for example. Since minimal technical skills and no knowledge of music theory are required to participate in CMM, it can be an ideal method of giving participants—children or adults—their initial experiences in the process of creating music.

USING CLASSROOM INSTRUMENTS

Any available musical instrument in the classroom can be utilized in a CMM improvisation—for example, the instruments of Carl Orff, rhythm instruments, recorders, and the classroom piano.

For the purposes of using CMM at the grade school level, it is recommended that the sessions be limited to small groups, with the number of participants determined by the age of the children. Younger children will do best in a very small group of three to six players, with the teacher/facilitator being one of the group. This will enable the facilitator to better keep everyone focused on their activities. It is also recommended the improvisations be fairly short—one or two minutes in duration—to accommodate the children's shorter attention spans. The ending of each improvisation can be determined arbitrarily by the facilitator, especially if and when the children's attention seems to be straying.

With adult CMM participants, small groups of players would be beneficial, too, but it would also be possible to include any nonplaying listeners in a rotation with the players, so that the class size could be larger. The improvised pieces could also be as long as the players want. As is standard practice in CMM, the improvisations should be recorded and then played back for listening and discussion. Again, shorter pieces will enable younger children to remain more attentive while listening.

In the classroom the job of dealing with a recording device becomes more challenging for the teacher/facilitator. The ideal scenario would be to have another adult operating the recording equipment. In most situations that might not be possible, so the teacher/facilitator will have to be thoroughly knowledgeable in working with the specific recording equipment at hand. The simpler the equipment, the better.

In the event of any technical problem, the teacher/facilitator must either solve the problem in short order, or stop trying to record altogether and save it for a future session. The bottom line is to maintain contact with the students and to keep the focus on the music making. As frustrating as small technical problems can be, it would be much worse to have students unfocused for any length of time while the facilitator is struggling with the equipment.

The questioning and discussion steps can also be brief for children: "Do you remember what you played? Did you like the music? Could you hear yourself?"

At younger ages, CMM will be most effective if there are only a few students with a facilitator and they are all engaged in the four CMM steps together. By the middle school ages, it ought to be possible, based on the teacher's assessment of the students' abilities, to have larger groups with rotation between performers and listeners, more closely resembling the CMM process presented in earlier chapters.

STANDARDS IN MUSIC EDUCATION

In some parts of North America, national, statewide, or provincial standards in music education provide that grade school students should have classroom experiences in music

Fig. 19.1 At younger ages, creative music making will be most effective if there are only a few students playing along with a facilitator.

improvisation. While educators may disagree about the effects of such standards on learning, the CMM process nevertheless clearly provides a method that can be used by teachers in meeting music standards.

The National Association for Music Education (MENC) publishes a document titled "The K–12 National Standards, Pre-K Standards, and What They Mean to Music Educators." It outlines standards in music for each of the various grade levels and provides guidelines for meeting those standards.

In the "Pre-K" (pre-kindergarten) section, there are four content standards:

1. Singing and playing instruments
2. Creating music
3. Responding to music
4. Understanding music

Each of the Pre-K standards is accompanied by several *achievement standards* that essentially describe specific activities that any child who meets the content standards ought to be able to accomplish. For example, under Pre-K content standard 2, one of the achievement standards is that "children [can] improvise instrumental accompaniments to songs, recorded selections, stories, and poems."

CMM, as a process, can usefully be applied to this Pre-K achievement standard. The facilitator/teacher could narrate a nursery rhyme and even play along while a child improvises on a classroom instrument or a homemade amadinda. The improvisation can be recorded and played back for listening. A few simple questions can follow: "Did you like that? Can you hear yourself? Did the music make you feel happy or sad?" Or, the teacher/facilitator can preface the questions with statements such as, "I liked that! Did you? I heard you playing! Did you hear it, too? That music made me smile! Did you smile, too?"

Because of the very short attention span of preschoolers, CMM would be most effective in a one-on-one environment, or possibly with two or three children at most. It is simply assumed that children at the prekindergarten age will not sit and listen while others are playing and recording. The teacher/facilitator will certainly need to have a sensitivity to the children's overall level of involvement, and the flexibility to change strategies in keeping them engaged.

In a similar format, the MENC national K–12 content standards are broken down into three grade levels:

1. Grades K to 4—eight content standards
2. Grades 5 to 8—nine content standards
3. Grades 9 to 12—nine content standards

Without attempting to go into every single MENC standard, it is enough to suggest here that the process of creative music making can be applied to many of the national achievement standards in music. More detailed information about MENC and the national standards can be accessed online at www.menc.org.

Similarly, the New York State Education Department (NYSED) publishes "Arts Learning Standards and Performance Indicators." This document is available online at www.emsc.nysed.gov/guides/arts. It contains four standards for music that are collectively intended to serve as general guides for grade school teachers.

Although there is some similarity with the MENC standards, the format is somewhat different. Each NYSED music standard is accompanied by a number of "indicators" that, if performed by students, would demonstrate that students are meeting the standard. The indicators are given for each of four grade levels—elementary, intermediate, commencement in general education (nonmusic major), and commencement majoring in music.

The NYSED arts standard 1 is called "Creating, Performing, and Participating in the Arts." To meet this standard in music, "students will compose original music and perform music written by others. They will understand and apply the basic elements of music in their performances and compositions. Students will engage in individual and group musical and music-related tasks, and will describe the various roles and means of creating, performing, recording, and producing music."

In order to attain the goals of the standards in music, the New York State School Music Association (NYSSMA) has developed a number of specific activities, or "tasks," that, if performed by students, would indicate that the goals have been met. One specific task recommended by NYSSMA for elementary school (and older) students is that elementary students "improvise a short composition using the black keys on a piano."

In CMM, a variation of this task would be possible: improvise a short composition using any of the keys on a piano and record it. Then listen to it and discuss it.

Another task suggested in the NYSED arts standard 1 at the intermediate grade level is for students to "improvise (on an instrument) a four-measure phrase that is coherent and expressive." CMM can provide an easy and workable method for students to perform this task and to build upon it.

USING A HOMEMADE AMADINDA

NYSED arts standard 2 is called "Knowing and Using Arts Materials and Resources." To meet this standard "students will use traditional instruments, electronic instruments, and a variety of nontraditional sound sources to create and perform music." Included among recommended tasks is that students "improvise a group composition."

The homemade amadinda, which was described in chapter 4, can be a valuable tool for use in the classroom in meeting this standard. Constructing an instrument in the classroom and then playing music on it can be an engaging classroom project, with obvious cross-disciplinary links between music, science (physics), social studies, and interpersonal (social) behavior.

The homemade amadinda—as an engaging and motivating musical instrument, especially when constructed by the students themselves—might well be helpful in meeting the requirements of state and national standards. Depending on the grade level, each homemade amadinda can accommodate as many as six players at a time. By utilizing the four steps of CMM, playing on a homemade amadinda can provide students of any age with meaningful experiences in improvising music.

AMADINDA GAMES

Because the vocabularies and skills of very young school children are likely to be minimal, various kinds of musical games can be helpful as simple developmental exercises while adding fun to the playing experience.

For example, one kind of game might be called Follow the Leader. Each player on one side of the homemade amadinda plays on a bar, and the player on the opposite side of the instrument must follow by playing the same note. Then, the sides may switch—the followers become the leaders, and vice versa.

Another kind of game might be called Boardwalk March. The players on one side of the amadinda must play on every regular beat, very much like a bass drum in a marching band. The players on the opposite side can play whatever rhythm and melody they want to. The roles may then be reversed.

LISTENING AND QUESTIONING

Regardless of whether a homemade amadinda or any other available classroom instruments are used for playing improvisations, the additional steps of recording, listening, and questioning are essential in the creative music making process. Again, there is a correlation with the NYSED standards.

Arts standard 3 of the NYSED is "Responding to and Analyzing Works of Art." In music it requires that "students will demonstrate the capacity to listen to and comment on music."

In an elementary classroom the CMM process can directly help to meet arts standard 3. In listening to their own music, students may be more engaged than they would be otherwise. In describing what they played and what they heard in playback, students may learn to recognize the elements of music—melody, rhythm, harmony, and the like. At the same time, they may learn about interacting with others—discovering what effects their actions have on others, and what effects the actions of others have upon them.

NYSED arts standard 4 is called "Understanding the Cultural Dimensions and Contributions of the Arts." To meet this standard in music, "students will develop a performing and listening repertoire of music of various genres, styles, and cultures that represent the peoples of the world and their manifestations in the United States."

Fig. 19.2 The questioning and discussion steps can be brief for children: "Did you like the music?" "What would you call this music?"

In CMM, it is supplemental listening that is used to expand the listener's musical vocabulary. Essentially, that is what the NYSED standard 4 in music is intended to accomplish, although in grade school classrooms the focus is more on appreciation and understanding than application in performance.

Again, it is not necessary to systematically make linkages among the many NYSED standards, the NYSSMA tasks, and the CMM process. The few brief examples given above should be enough to demonstrate that creative music making can be a useful pedagogy for grade school music educators. More information about the New York standards may be found online or by contacting the New York State Education Department at the address given in the bibliography.

20

OUTCOMES

I asked myself a question: What do I want my students to be able to do? And I thought, I want them to be able to work independently and know what they are doing.

—Dorothy DeLay, quoted in Barbara Lourie Sand, *Teaching Genius: Dorothy DeLay and the Making of a Musician*

As suggested earlier, one of the desirable outcomes of creative music making sessions would be the desire of participants to actually perform free-form improvisations in public concerts. In fact, it is recommended that a public performance be planned as the culminating event of either a CMM course or workshop. For this kind of concert, a variety of ensemble configurations—duets, trios, quartets, or larger—can be utilized. Varying combinations of instruments—strings, winds, percussion, or a mix—can also be used. Participants might form their own groupings or ensemble members may be appointed by the facilitator, or some combination of the two methods may be used.

Just as in CMM sessions, in a public performance, novice participants may be paired with professional musicians to the benefit of both. The novice is challenged to rise to the musicianship of the professional, and the professional, rather than being obliged to demonstrate technical mastery, is instead challenged to create an effective communication with the novice, a task requiring well-developed listening skills and imaginative responses. What better way is there for any professional to broaden perspective, to get beyond confining concepts of music making, to embrace a global rather than a specialized view? The shared goal is to make music.

HELPING LISTENERS IN A CONCERT AUDIENCE

I read this somewhere—that mystery is not the absence of meaning, it's the presence of more meaning than you can possibly comprehend.

—Bob Becker of NEXUS, interviewed by the author in 1998

Helping audiences to have a meaningful listening experience is an important responsibility for performing musicians in today's musical environment. A printed program with performance notes is the traditional method of conveying background information about composers and compositions.

While many musicians are of the opinion that audiences should never be addressed from the stage because it is disrespectful, many people—especially those living outside of the sophisticated metropolitan arts centers—appreciate commentary by performers, provided it is not too long or technical, which can indeed be intimidating and therefore counterproductive.

However, it is reasonable to assume that most audiences have little or no experience in listening to free-form improvisations and will probably benefit from a more personal kind of communication. The intent is to set the stage for a more involved and less intimidating listening experience. This can be accomplished by simply having one (or more) of the performers speak to the audience just before performing the improvisation:

> *You are about to hear music that only we here in this auditorium will ever be able to experience. It will be created especially for you, never to be heard again [unless, of course, it is recorded]. The music, which we have titled "___________," is a free-form improvisation, created entirely on the spot by the musicians here on stage without any preplanned structure. There are only two rules that will guide us: we can play whatever we want to play at any time, and we will listen—to each other and to ourselves; but there is no penalty if somebody doesn't listen. Our challenge is not to improvise the equivalent of a Beethoven symphony; it is to make music that is interesting for you and us to hear. We invite you to listen along with us as we go on our musical exploration.*

21

IMPROVISATION IN CONCERTS

> *Creative persons differ from one another in a variety of ways, but in one respect they are unanimous: They all love what they do.*
>
> —Mihaly Csikszentmihalyi, *Creativity: Flow and the Psychology of Discovery and Invention*

Creative music making, as a pedagogy, was developed originally for use in the context of an educational environment—in music schools or workshops—for the purpose of helping participants to broaden their musicianship in whatever genre of music to which they happen to be committed.

Free-form improvisation comprises the first step of CMM, but it is certainly possible for professional musicians to incorporate it into their activities outside of any formal education environment—for example, in public concerts presented before paying audiences.

One of the many wonderful qualities of free-form improvisation is that it can be appropriate for so many live performance situations, from chamber music concerts to jazz festivals to world music events. It is a natural "crossover" vehicle, even though the idea of making a crossover connection may be the last thing on the minds of free-form improvisers.

To illustrate the variety of possibilities, NEXUS has performed free-form improvisations with musicians and ensembles as diverse as Abraham Adzenyah (Ghanaian master drummer); Earle Birney (Canadian poet); the Chautauqua, New York, Symphony Orchestra; David Darling (cellist); Peter Erskine (drummer); Steve Gadd (drummer); the Hamilton, Ontario, Philharmonic Orchestra; Paul Horn (flautist and saxophonist); the Kronos Quartet (string quartet); Tony Levin (stick player/bassist); Oscar Peterson (pianist); and Richard Stoltzman (clarinetist).

Free-form improvisations with these artists have involved the full spectrum of time lengths depending on each particular circumstance—everything from a single five-minute improvised piece in the context of a solo concert to a full concert program of free-form improvisations.

One particular concert performed by NEXUS on the University of Toronto's Faculty of Music Chamber Music Series provides a good example for a closer look at the possibilities offered by programming free-form improvisations. The concert occurred on November 17, 2003, and the entire first half of the program consisted of two improvised pieces.

The opening improvisation was titled "Sound Sculpture." It was performed by NEXUS using a variety of world percussion instruments belonging to the individual members of the ensemble. Filling the stage were drums, cymbals, a marimba and vibraphone, racks of suspended bells and gongs, and tables covered with wood blocks, shakers, animal bells, African percussion instruments, and tools for making assorted sound effects. There were also several kinds of harmonicas, whistles, a digital sampler, and a small group of unique sound sculptures designed and made in France by François Baschet.

An important aspect of NEXUS improvisations over the years has been that the instruments used are almost always different for each concert, depending on the preferences of each individual player. The result has been that the orchestration is different each time, which has brought a sense of freshness to each improvisation. The regular addition of newly acquired instruments by each individual player from time to time has also provided a constant stream of new sounds, which has enabled the improvisations to continually maintain a sense of discovery.

In "Sound Sculpture," the Baschet instruments were played in a NEXUS improvisation for the first time. These instruments, owned by NEXUS member Garry Kvistad, are mostly metal rod constructions. On each instrument's rod frame there are combinations of plastic resonating cones, glass rods, metal rods, metal bars, and springs that can be plucked, struck with mallets, or stroked with a damp cloth. The ability to include such instruments—sculptures, homemade or "found" instruments—only adds to the possibilities for musical exploration and discovery in free-form improvisation.

The second improvisation on the concert was titled "Nimmons 'n NEXUS." It featured clarinetist Phil Nimmons, who has taught with the faculty of music at the University of Toronto for over thirty years and who is known—especially throughout Canada—as a performer, composer, educator, music clinician, and artistic director of music programs.

Play CD Track #13—Nimmons 'n NEXUS (excerpt) (Time: 12:16)

Players: Phil Nimmons, clarinet; NEXUS—Bob Becker, Bill Cahn, Robin Engelman, Russell Hartenberger, and Garry Kvistad

Phil Nimmons and NEXUS were no strangers to each other before the concert. We had previously performed together a number of times in free-form improvisations that had always been structured in the same way—the only plan was an unspoken one, that each player could play whatever he wanted to play. It was simply assumed, but never specifically stated, that every player, in accordance with his own musicianship and sensibilities, would also be listening to everything and playing in response to whatever was heard. However, it was also understood that in the event one of the players, for whatever reason, should stop listening carefully, he would almost certainly be the only one to ever know it.

The instruments used by NEXUS were essentially the same as in "Sound Sculpture." However, one feature of this improvisation was the extended use of a digital sampler. One of the interesting aspects of NEXUS improvisations has always been the ensemble's ability to explore acoustic sounds on their instruments—sounds that were frequently mistaken by

Fig. 21.1 Phil Nimmons, clarinet, and NEXUS member Garry Kvistad playing Baschet sound sculptures.

listeners to be electronic sounds. The harmonic wind chimes (The Woodstock Synergy Chime™) heard in "Nimmons 'n NEXUS" provide a good example.

In this improvisation, the digital sampler contained voicings sampled from acoustic instruments belonging to the players—mainly bells and gongs—with the intent of extending the ranges of the acoustic instrument sounds from which the samples had been made. In this scenario the listener is challenged to distinguish between the electronic and acoustic sounds, because the sampled sounds can easily be mistaken for acoustic ones, even by the players.

The performance of "Nimmons 'n NEXUS" proved to be very satisfying for the players as well as for the audience. In the review of the concert that was published in the Toronto *Star*, music critic William Littler commented that "there has always been a wonderful sense of connectedness in the way these [NEXUS] musicians improvise together, seeming to inhabit each other's thought processes … it proved equally interesting to hear how wind and percussion could take inspiration from each other and undergo subtle personality changes in the process."

It is worth noting that the newspaper review is largely concerned with the interaction of the players rather than with the musical material. It is a recognition, consciously or not, that it is in the nature of free-form improvisation for a central focus to be on the actions and responses of the people involved—the players and the listeners—rather than on the compositional structure.

As for the music itself, the review was consistent with the kinds of responses normally received by NEXUS after free-form improvisations. Listeners don't refer to wrong notes;

there were none, nor could there have been any. They don't analyze the structure or form of the improvised composition; there was a structure, but that was not the point of the music. They don't offer compliments about the virtuosity of the players; in fact, if listeners are thinking about a player's technical difficulties, their attention has been misdirected by the player.

Instead, audiences frequently comment on the way the players work together to give the music an organic quality. Listeners certainly can hear and be fascinated by the unique and unfamiliar sound combinations that have been produced in an improvisation, but the sounds primarily serve as the vehicle by which performers and listeners make connections and create contrasts.

Such responses are not unlike the responses generated in creative music making sessions. Both the CMM process and the public performance of free-form improvisations encourage performers and listeners to broaden their modes of thought and expand their repertoire of possibilities.

Fig. 21.2 "Nimmons 'n NEXUS," an improvisation performed by (left to right) Bill Cahn, Bob Becker, Robin Engelman, Russell Hartenberger, Phil Nimmons, and Garry Kvistad.

BUILDING A HOMEMADE AMADINDA

MATERIALS NEEDED

Four 8-foot lengths of 2" x 4" pine or cedar (knots are okay)
One 3-foot length of 2" x 4" pine or cedar (knots are okay)
One square yard of $\frac{1}{2}$"-thick poly foam or outdoor carpeting
One 5-foot length of $\frac{1}{4}$"-diameter pine dowel
Two 4-foot lengths of $\frac{1}{2}$"-diameter pine dowel
One 4-foot length of neoprene rubber tubing ($\frac{1}{2}$" inner diameter, $\frac{5}{8}$" outer diameter)

TOOLS NEEDED

A power saw or hand saw
A power drill with a $\frac{1}{4}$" bit and a $\frac{5}{16}$" bit
A pair of heavy scissors (or a utility knife)
A bottle of wood glue
A lead pencil

DIRECTIONS

1. Using proper safety precautions, saw the 2" x 4" pine or cedar keys to the following lengths:
   8-foot board #1—45", 30", 18"
   8-foot board #2—41", 37", 15"
   8-foot board #3—60", 26"
   8-foot board #4—60", 22"
   3-foot board—33"
2. File and sand smooth all surfaces, corners, and edges.
3. Drill two $\frac{1}{4}$" holes in the 45" (largest) key; each hole will be 9" in from either end and centered between the sides of the top playing surface.
4. Drill two $\frac{1}{4}$" holes in the 15" (smallest) key; each hole will be 3" in from either end and centered between the sides of the top playing surface.
5. Place all of the keys on the rails; space the keys $\frac{1}{2}$" apart; the rails should lay directly beneath the two holes drilled in both the largest and smallest keys.

6. On the top of both rails make pencil lines along the sides of all keys.
7. On each key make a pencil mark *X* at the center point of the key where it rests over each rail. (Each key will have 2 pencil marks.)
8. Saw the $\frac{1}{4}$" dowel into eighteen 3" lengths; sand both ends of each piece smooth.
9. Drill two $\frac{1}{4}$" holes (approx. $1\frac{1}{2}$" deep) into the top of each rail, through the predrilled holes in the largest (45") and smallest (15") keys.
10. Insert and glue 3"-long $\frac{1}{4}$" dowels into the four rail holes; set aside the largest and smallest keys (i.e., do not lay the keys in over the dowels yet).
11. Align the remaining eight keys to the pencil lines on the rails; one by one, directly over each rail, drill a $\frac{1}{4}$" hole through the key and into the rail $1\frac{1}{2}$" deep; set aside each key in turn and insert and glue 3"-long dowels into the rail holes; repeat until all keys are drilled and dowels inserted into both rails.
12. Cut the poly foam or carpeting into $1\frac{1}{2}$"-wide strips; allowing a $\frac{1}{2}$" overhang at each end of the rails, lay the insulating strips end-by-end over each rail and then, using the scissors, punch holes in the strips where the dowels occur.
13. In all nine keys redrill both holes to make the holes $\frac{5}{16}$" in diameter; place all keys on the rails; the holes in the keys should fit loosely over the dowels.
14. Saw the $\frac{1}{2}$" dowel into 8" lengths; cut the neoprene rubber tubing into 4" lengths.
15. Place the neoprene rubber tubing over one end of each 8" dowel mallet.
16. Use the rubber end of the mallet and strike the top corner at the end of each key to play the sound. (The mallet will be at an angle.)

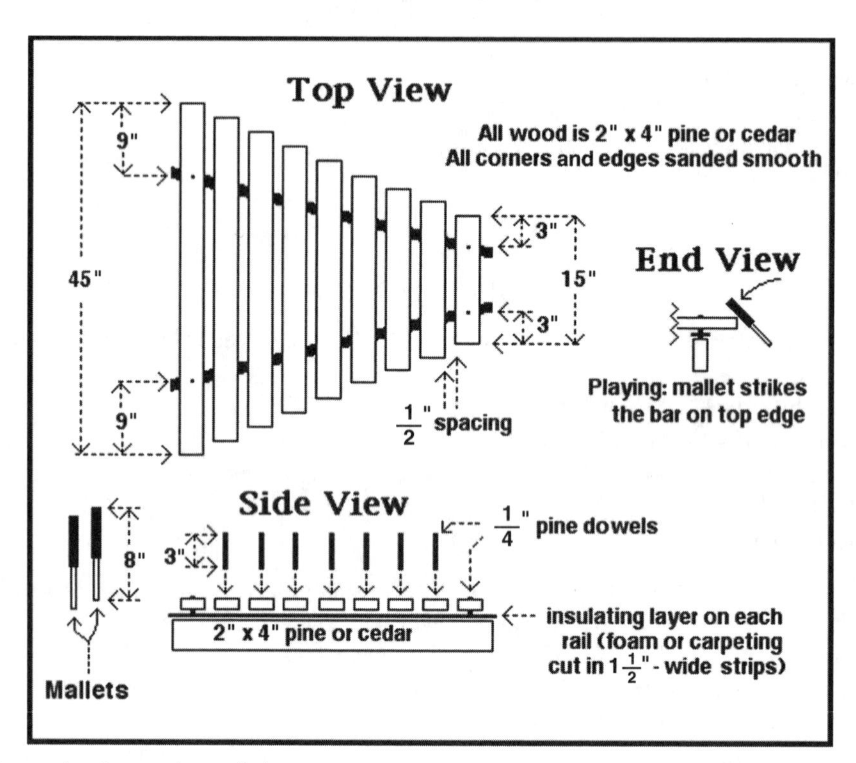

Fig. A.1 Diagram for a homemade amadinda.

EXAMPLES FOR SUPPLEMENTARY LISTENING

The following is a list of musical examples that illustrates the kinds of music that might be selected for a supplemental listening session. Of course, any music may be selected. This list intentionally includes a diversity of musical styles and aesthetics. The goal is to provide creative music making participants access to a wide spectrum of music and vocabularies in order to stimulate questioning and dialogue.

1. "Light Up My Room" (music: Ed Robertson/words: Ed Robertson and Steven Page)
   Barenaked Ladies: Jim Creeggan, double bass; Kevin Hearn, keyboards/vocal; Steven Page, acoustic guitar/vocal; Ed Robertson, lead vocal/acoustic and electric guitars; Stewart Tyler, drums
   (*Stunt*, Reprise CD no. 9 46963-2)
   Time: 3:37
2. *Piano Concerto No. 27 in B-flat Major*, K. 595 (2. Larghetto) (music: Wolfgang Amadeus Mozart)
   Jacques Abram, piano; with the Vienna Chamber Orchestra, Helmuth Froschauer, conductor
   (Musical Heritage Society LP no. 831)
   Time: 7:59
3. "Dance of the Octopus" (music: Red Norvo)
   Red Norvo, marimba; Benny Goodman, bass clarinet; Artie Bernstein; bass, Dick McDonough, guitar
   (*Red Norvo & His All Stars*, Epic LP no. LG3128)
   Time: 3:35
4. *Nocturnes* (2. Fêtes) (music: Claude Debussy)
   London Symphony, Leopold Stowkowski, conductor
   (Capitol LP no. P8520)
   Time: 6:30
5. "Drowsey Maggie" (music: Irish traditional)
   The Chieftains: Michael Tubridy, flute; Seán Potts, tin whistle; Paddy Moloney, Uilleann pipes; Martin Fay, fiddle; Peadar Mercier, bodhrán and bones
   (*The Chieftains 4*, Island LP no. ILPS 9380)
   Time: 4:15

6. "Elysium" (excerpt) (music: Branford Marsalis)
   Branford Marsalis Contemporary Jazz: Branford Marsalis, saxophone; Joey Caldero, piano; Jeff "Tain" Watts, drums; Eric Revis, bass
   (*Contemporary Jazz*, Columbia CD no. CK63850)
   Time: 2:10
7. "Cérémonial des Janissaries" (music: Turkish traditional)
   L'Ensemble de l'Armé de la République Turke, Kudsi Erguner, director
   (*Les Janissaires*, Ethnic CD no. B 6738)
   Time: 2:10
8. "Gagaku" (music: Japanese traditional)
   Gagaku Ensemble
   (*Hogaku Excellent*, Columbia audiocassette no. CAY 9038)
   Time: 13:00
9. "Lagu Kodok" (Frog Song) (music: Balinese traditional)
   Genggong Batur Sari, Batuan
   (*Bali*, Elektra/Nunsuch audiocassette no. 9 79204-4)
   Time: 4:11
10. "Didjeridu" (music: Australian Aboriginal traditional)
    Borroloola, Northwest Territory
    (*Aboriginal Sound Instruments*, Australian Institute of Aboriginal Studies audiocassette no. 14)
    Time: 1:00
11. "XpiotóöouAoç" (music: Greek traditional)
    Florina Brass Band
    (*Florina Brass Band*, Ano Kato audiocassette no. 2008)
    Time: 3:48
12. "Epitaph for Moonlight" (music: R. Murray Schafer)
    Amadeus Choir and NEXUS, Weston Recital Hall, Toronto, June 6, 1999
    Time: 5:35

RECORDINGS OF IMPROVISED MUSIC PERFORMED BY BILL CAHN

LPs

1. Paul Horn with NEXUS, *Paul Horn + NEXUS* (Epic LP no. KE-33581), 1975.
   tracks: "Somba," "Crystals," "Friendship," "NEXUS," "Mbira," "Latin Tala," "African Funeral Song," "Eastern Star," "Dharma," "Capetown."
2. Paul Winter, *Common Ground* (A&M LP no. SP-4698), 1978.
   track: "Eagle."
3. Paul Winter, *Callings* (Living Music LP no. LMR-1), 1980.
   tracks: "Love Swim," "Sea Storm," "Callings," "Sea Joy," "Talking Bells."
4. NEXUS, *Music of NEXUS* (NEXUS LP #NE-01), 1981.
   tracks: "Unexpected Pleasures," "Passage," "Amazing Space."
5. NEXUS and Earle Birney, *NEXUS and Earle Birney*, vols. 1–3 (NEXUS LP nos. NE-02, NE-03, NE-04), 1981.
   tracks: various.

CDs

1. Paul Horn with NEXUS, *Altitude of the Sun* (Black Sun CD no. 15002-2), 1989.
   tracks: "Somba," "Friendship," "NEXUS," "Mbira," "Latin Tala," "African Funeral Song," "Eastern Star," "Dharma," "Capetown."
2. NEXUS, *The Best of NEXUS* (NEXUS CD no. 10251), 1989.
   tracks: "Unexpected Pleasures," "Passage," "Amazing Space."
3. NEXUS, *Origins* (NEXUS CD no. 10295), 1992.
   tracks: "Origins," "Signs of the Time," "Song of the Nine Iron," "Perpetual Emotion," "Arioso," "Waltz Happens," "Our Way."
4. NEXUS with the Rochester Philharmonic Orchestra, Peter Bay, conductor, *Voices* (NEXUS CD no. 10317), 1994.
   tracks: "Connexus," "Voices" (some sections improvised).
5. Tony Levin with NEXUS, *World Diary* (Papa Bear CD), 1995.
   tracks: "Heat," "Expresso and the Bed of Nails."
6. NEXUS with Eric Robertson, *Toccata* (NEXUS CD no. 10410), 1997.
   tracks: "Kichari," "Reunion," "Toccata."

7. NEXUS with the Pacific Symphony, Carl St. Clair, conductor, *Takemitsu Orchestra Works* (SONY Classical CD no. SK 63044), 1998.
   tracks: "From Me Flows What You Call Time" (some sections improvised).
8. NEXUS with Richard Stoltzman, *Garden of Sounds* (BIS CD no. 1108), 2000. tracks: "Le Dialog du bois," "Eternal Triangle Beckons," "Wonderings," "Rhapsody in Green," "Rites," "Waterfall Having No Water," "Bamboo Grove," "Midnight in the Lupines," "Amazin' Gazin'," "Garden of Sounds," "Tibet to Be," "Reflections," "Ebony Reverie."

TRACK LIST FOR THE ENCLOSED COMPACT DISC

Track #1
Improvisation Duet—Eastman School of Music, July 19, 2002
Seung Hae Jung, piano; Ko Taniguchi, violin
Time: 5:21

Track #2
Improvisation Quartet—Eastman School of Music, July 19, 2002
Ya-ting Lee, piano; Edith Mann, percussion; Jin Kim Soo, violin; Juei-hsien Wang, piano
Time: 6:57

Track #3
Amadina Improvisation—Eastman School of Music, July 19, 2002
Bevin Coggeshall, Sebastian Henshaw, Edith Resnick, Jordan Schifino (high school participants in "Improvisation for All")
Time: 7:07

Track #4
Duet Improvisation—University of Missouri–Kansas City
Bill Cahn, vibraphone; Nick Urbom, marimba
Time: 4:31

Track #5
Trio Improvisation A— University of Missouri–Kansas City
Will Braune, percussion; John Thieben, saxophone; Sam Wisman, percussion
Time: 2:58

Track #6
Trio Improvisation B—University of Missouri–Kansas City
Roger Caliman, percussion; Hunter Long, saxophone; John Hillan, clarinet
Time: 2:29

Track #7
Trio Improvisation C—University of Missouri–Kansas City
Roger Caliman, percussion; Hunter Long, saxophone; John Hillan, clarinet
Time: 1:51

Track #8
Improvisation D—University of Missouri–Kansas City
Brenden Bennett, marimba; Liz Stephens, vibraphone; Ryan Wurtz, guitar
Time: 3:51

Track #9
Improvisation E—University of Missouri–Kansas City
Brenden Bennett, marimba; Liz Stephens, vibraphone; Ryan Wurtz, guitar
Time: 2:55

Track #10
Improvisation F—University of Missouri–Kansas City
Brett Baxter, percussion; Tom Kernan, percussion; Bill Solomon, piano
Time: 3:48

Track #11
Improvisation G—University of Missouri–Kansas City
Brett Baxter, percussion; Tom Kernan, percussion; Bill Solomon, piano
Time: 2:39

Track #12
Expanding Quartet Improvisation—University of Missouri–Kansas City
Group 1—four percussion players start
Group 2—four percussion players join in
Group 3—three players (saxophone, two percussion players) join in
Time: 16:16

Track #13
Nimmons 'n NEXUS (excerpt)
Phil Nimmons, clarinet; NEXUS—Bob Becker, Bill Cahn, Robin Engelman, Russell Hartenberger, and Garry Kvistad
Time: 12:21

Total Time: 73:13

BIBLIOGRAPHY

Atkins, Greg. *Improv! A Handbook for the Actor.* Portsmouth, NH: Heinemann, 1994.

Bailey, Derek. *Improvisation: Its Nature and Practice in Music.* New York: Da Capo, 1993.

Barzun, Jacques. *From Dawn to Decadence.* New York: HarperCollins, 2000.

Bergen, Mark, Molly Cox, and Jim Detmar. *Improvise This!* New York: Hyperion, 2002.

Boldt, Laurence G. *Zen and the Art of Making a Living.* New York: Penguin, 1993.

Booth, Eric. *The Everyday Work of Art.* Naperville, IL: Sourcebooks, 1999.

Blum, Stephen. "Recognizing Improvisation." In *In the Course of Performance*, ed. Bruno Nettl with Melinda Russell. Chicago: University of Chicago Press, 1998.

Cage, John. *M: Writings '67–'72.* Middletown, CT: Wesleyan University Press, 1972.

Cage, John. *Notations.* West Glover, VT: Something Else, 1969.

———. *Silence.* Cambridge, MA: MIT Press, 1961.

Cassirer, Ernst. *An Essay on Man.* New Haven, CT: Yale University Press, 1944.

Claxton, Guy. *Hare Brain Tortoise Mind: How Intelligence Increases When You Think Less.* HarperCollins, 2000.

Copland, Aaron. *What to Listen for in Music.* New York: McGraw-Hill, 1957.

Crofton, Ian and Donald Fraser. *A Dictionary of Musical Quotations.* New York: Schirmer, 1985.

Csikszentmihalyi, Mihaly. *Creativity: Flow and the Psychology of Discovery and Invention.* New York: HarperCollins, 1996.

Farson, Richard and Ralph Keyes. *Whoever Makes the Most Mistakes Wins.* New York: Free Press, 2002.

Gablik, Suzi. *Conversations Before the End of Time.* London: Thames and Hudson, 1995.

Gardner, Howard. *Creating Minds.* New York: Basic, 1993.

Gelb, Michael J. *How to Think Like Leonardo da Vinci.* New York: Delacorte Press, 1998.

Gershwin, George. "I Got Rhythm" from *Crazy Girl.* Los Angeles: W.B. Music Corp., 1930.

Gordon, Edwin E. *Rhythm.* Chicago: GIA, 2000.

Huxley, Aldous. *Music at Night.* London: Chatto & Windus, 1931.

Jaworski, Joseph. *Synchronicity.* San Francisco: Barrett-Koehler, 1996.

Kimball, Kathleen. *The Music Lover's Quotation Book.* Toronto: Sound and Vision, 1990.

Levesque, Lynne C. *Breakthrough Creativity.* Palo Alto: Davies-Black, 2001.

Littler, William. "Nexus Drums up a Worldwide Reputation." *Toronto Star*, March 7, 1982.

Littler, William. "Nexus Elevates Fine Art of Banging on Things." *Toronto Star*, November 1, 2003.

Mathieu, W. A. *The Listening Book: Discovering Your Own Music.* Boston: Shambhala, 1991.

May, Rollo. *The Courage to Create.* New York: W. W. Norton, 1975.

MENC: The National Association for Music Education, *The K–12 National Standards, Pre-K Standards, and What They Mean to Music Educators.* Reston, VA: MENC, 2000; www.menc.org.

Menuhin, Yehudi. *Theme and Variations.* Oxford: Heinemann Educational, 1972.

Nachmanovich, Stephen. *Free Play: Improvisation in Life and Art.* New York: Penguin Putnam, 1990.

Nettl, Bruno. "An Art Neglected in Scholarship." In *In the Course of Performance*, ed. Bruno Nettl with Melinda Russell. Chicago: University of Chicago Press, 1998.

Nettl, Bruno with Melinda Russell, eds. *In the Course of Performance.* Chicago: University of Chicago Press, 1998.

New York State Education Department. *Arts Learning Standards and Performance Indicators.* Albany, NY: NYSED, 2004; 89 Washington Ave., Albany, NY 12234; www.emsc.nysed.gov/guides/arts.

Pleasants, Henry. *The Agony of Modern Music.* New York: Touchstone, 1955.

Pressing, Jeff. "Psychological Constraints on Improvisational Expertise and Communication." In *In the Course of Performance*, ed. Bruno Nettl with Melinda Russell. Chicago: University of Chicago Press, 1998.
Price, Theodore. "International Art, Meditative Music." Rochester (NY) *Democrat & Chronicle*, May 23, 1971.
Reich, Steve. *Writings about Music*. New York: New York University Press, 1974.
Sand, Barbara Lourie. *Teaching Genius: Dorothy DeLay and the Making of a Musician*. Portland, OR: Amadeus, 2000.
Saul, John Rolston. *On Equilibrium*. Toronto: Penguin Canada, 2001.
Shapiro, Nat. *An Encyclopedia of Quotations about Music*. New York: Da Capo, 1977.
Webster's New School and Office Dictionary. Cleveland and New York: The World Publishing Company, 1960.
Woodruff, Paul. *Reverence*. New York: Oxford University Press, 2001.
Wyre, John. *Touched by Sound*. St. John's, Newfoundland: Buka Music, 2002.

INDEX

A

Amadinda
  Amadinda Improvisation (CD Track #3), 16, 109
  description of, 16
    games using, 94–95
    homemade, 94, 103–104
Amateur musicians, 15
Art music, 26–27
Atkins, Greg, 28
Audiences, 98, 102

B

Bailey, Derek, 31
Bechet, Sidney, 24
Becker, Bob, 1, 4, 97, 100
Bells, 1–2
Benson, Warren, 3
Blum, Stephen, 25
Boldt, Laurence, 42
Booth, Eric, 7
Busoni, Ferruccio, 40

C

Cage, John, 2
Career-track music students, 13–14
Cassirer, Ernst, 1
Chaotic music, 28
Clap/change exercise, 43
Classical music
  improvisation in, 12, 26–27
  musical elements in, 26
  orchestral, 26
Claxton, Guy, 21
CMM, *See* Creative music making
Coggeshall, Bevin, 16
Commercial use of recording, 48
Competition, 11
Concerts, 99–102
Consonance, 2, 42, 73
Copland, Aaron, 57
Copyright, 46–47
Craden, Michael, 40
Creative act, 23
Creative music making
  accessibility to, 29
  benefits of, 13–14, 38–39
  definition of, 12–13, 19
  environment for, 35–36, 58
  exercises in, 42–43
  facilitator of, *See* Facilitator
  goals of, 9, 29, 33
  in grade schools, 91–92
  instruments used in, 13, 91–92
  listening step of, *See* Listening
  outcomes of, 29, 97–98
  as pedagogy, 91, 99
  playing step of, *See* Playing
  practice in, 33
  practitioners influenced by, 13–17
  questioning step of, *See* Questioning
  questions associated with, 7
  recommendations for, 31
  recording step of, *See* Recording
  rules for, 20
  self-imitation exercises, 42–43
  as soloist, 43–44
  steps involved in, 19, 31
  as study course, 34
  technology influences, 15
  as workshop, 34
Creative music making sessions
  description of, 33–34

outcomes of, 97–98
at University of Missouri–Kansas City Conservatory of Music, *See* University of Missouri–Kansas City Conservatory of Music sessions
Creativity, 20–21, 27
Csikszentmihalyi, Mihaly, 99

D

Debussy, Claude, 31
DeLay, Dorothy, 97
Disengagement from music, 41
Drum circles, 44
Duet Improvisation (CD Track #4), 61–63

E

Engelman, Robin, 1, 4
Ensemble, 35–36
Environment, 35–36, 58
Expanding Quartet Improvisation (CD Track #12), 85–89

F

Facilitator
in ensemble, 36
environment created by, 37
in grade school settings, 92
in playing step, 32, 37, 39
in questioning step, 54
responsibilities of, 33
role of, 29–30, 32, 39
trust established by, 32
Families, 17
Free-form improvisation, *See also* Creative music making
accessibility of, 28
aesthetics of, 28–29
beauty in, 30
benefits of, 14
characteristics of, 27–28
in concerts, 99–102
creativity in, 27
description of, 5, 12, 99
ending of music in, 41
experience in, 28
imitative playing in, 42
musical ideas in, 49
objectives of, 30
performance options during, 41
premise of, 30
qualities of, 99
as quasi-composition, 42
reactions to, 28–29
"right" and "wrong" in, 28
rules for, 20, 41–42
self-imposed rules in, 41–42
supplemental playing rules, 42

G

Gelb, Michael J., 53
Gershwin, George, 44
Glockenspiels, 16
Gordon, Edwin E., 49, 56
Grade schools
creative music making in, 91–92
listening in, 95–96
music education standards in, 92–94
questioning in, 92, 95–96
Grading, 11
Group tag-team improvisation, 43

H

Hartenberger, Russell, 1–2, 4
Haydn, Franz Joseph, 26
Henshaw, Sebastian, 16
Huxley, Aldous, 45

I

Idiomatic improvisation
definition of, 24
in drum circles, 44
in jazz, 24–25
in non-western music, 25–26
stylistic rules of, 24
in western art music, 26–27
in western classical music, 26–27
Imitation
as an exercise, 43
self-imitation, 42–43
Imitative playing, 42
Improvisation
abstract nature of, 5
abstract visual, 27
in acting, 24
boundaries of, 12
in classical music, 12
in concerts, 99–102
definition of, 23
description of, 1, 3–4
in drum circles, 44

examples of, 12, 38
forms of, 24
free-form, *See* Free-form improvisation
in grade schools, 91–92
graphic, 27
group tag-team, 43
idiomatic, *See* Idiomatic improvisation
nonidiomatic, 27
in nonmusical activities, 24
pedagogical uses of, 5–6
pedagogy of, 12
playback of, 38, 50
playing of, *See* Playing
recording of, *See* Recording
solo, 43–44
tag-team, 43
teaching uses of, 5–6
in theater arts, 24
workshops, 4
Improvisation Duet (CD Track #1), 14, 109
Improvisation Quartet (CD Track #2), 14, 109
Instruments, 13, 91–92
Intuition, 21–23

J

Jaworski, Joseph, 51, 55
Jazz
genres of, 25
idiomatic improvisation in, 24–25
musical elements in, 25

K

Keyboard instruments, 15
Kilbourn Hall, 3
Kodaly, Zoltan, 91
Kumar, Satish, 19

L

Landowska, Wanda, 33
Learning in performing music, 7
Lee, Ya-Ting, 14
Levesque, Lynne C., 20
Listening
audience participation in, 98
description of, 19, 35
in grade school settings, 95–96
importance of, 32, 49
to improvisation playback, 50
involvement in, 51
objective, 51, 57
playing and, 49
subjective, 51
supplemental, *See* Supplemental listening
techniques used in, 50
vocabulary building through, 51–52

M

Ma, Yo-Yo, 26
Mann, Edith, 14
Marimba, 5, 15
Mathieu, W. A., 50
McLuhan, Marshal, 51
Meaning in music, 57–58
Menuhin, Yehudi, 23
Mistakes, 37
Mixed instrumentation, 59
Music
classical, *See* Classical music
disengagement from, 41
ending of, in free-form improvisation, 41
impact of, 8
keeping in touch with, 40–41
learning about, 7
meaning in, 57–58
"representational" connection to, 56–57
understanding of, 56–57
valuing of, 56
Music education
competition influences on, 11
emphasis shifts in, 8
focus of, 13–14
grading in, 11
historical changes in, 8–9
standards in, 92–94
technological influences on, 8
Music making
creative, *See* Creative music making
objectification of, 12
Music performances
audiences to, 98, 102
during free-form improvisation, 41
"fixing" of, 21–22
public, 97
reviews of, 101–102
"studied," 22
technological influences on, 9
Music students
career-track, 13–14
creative music making effects on, 13–14
general, 15
Music teachers, 14
Musical ideas, 49
Musical vocabulary, 51–52

Musicians
  amateur, 15
  novice, 15–16
  professional, 14
Musicianship
  elements of, 11
  pedagogy in, 12
  subjective nature of, 12

N

Nachmanovich, Stephen, 34
National Association for Music Education, 93
Nettl, Bruno, 27
New York State Education Department music standards, 93–94
New York State Music Association, 94
NEXUS, 1–4, 99–101
Nimmons 'n NEXUS (CD Track #13), 101
Nimmons, Phil, 100–101
Nonidiomatic improvisation, 27
Nonmusicians, 17
Noun music, 28
Novice musicians, 15–16

O

Objective listening, 51, 57
Ostinato, 40, 44
Outcomes, 97–98

P

Participants
  mimicking by, 33
  relaxation by, 33
  trust of, 32
Percussion instruments, 1, 3–4
Performances, *See* Music performances
Piano, 15
Playback
  description of, 38
  listening to, 38, 55
  questions after, 55
Playing
  description of, 19, 21
  ensemble size, 35–36
  facilitator's role in, 32
  imitative, 42
  keeping in touch with music during, 40–41
  listening and, 38, 49
  mistakes during, 37
  objective of, 32
  phase 1, 37
  phase 2, 38
  phase 3, 38–39
  questions after, 53–54
  range of, 39
  recording step and, 45
  rhythmic groove during, 40
  rules for, 35
Pleasants, Henry, 8
Practice, 33
Pressing, Jeff, 13
Professional musicians, 14
Promotional use of recording, 46–47
Public performances, 97

Q

Questioning
  after listening to playback, 55
  after playing step, 53–54
  definition of, 53
  description of, 19
  facilitator's role in, 54
  in grade school settings, 92, 95–96
  importance of, 32, 53

R

Recording(s)
  announcement of players' names, 46
  CD made during, 46–47
  for commercial purposes, 48
  description of, 5, 19
  for educational purposes, 46
  equipment needed for, 45–46
  importance of, 45
  list of, 107–108
  playing step and, 45
  for promotional purposes, 46–47
  recording equipment, 45–46
  transcription from, 48
Reich, Steve, 55
Relaxation, 33
"Representational" connection, 56–57
Resnick, Edith, 16
Rhythmic groove, 40, 44
Rzewski, Frederic, 4

S

Santur, 25
Sarangi, 25
Schifino, Jordan, 16

Self-imitation, 42–43
Sessions, *See* Creative music making sessions
Seung Hae Jung, 14
Solo improvisation, 43–44
"Sound Sculpture," 100
Spirituality, 9
Spontaneity, 22
Starr, Ringo, 29
Subjective listening, 51
Supplemental listening
  description of, 51–52
  examples for, 105–106
  in grade school settings, 96
  questions after, 55–56
Synchronicity, 42

T

Tag-team improvisation, 43
Takemitsu, Toru, 28
Taniguchi, Ko, 14
Technical evaluation, 11
Technology
  for creative music making, 15
  music education affected by, 8
  music performances affected by, 9
"Time," 40
Transcription of recording, 48
Trio Improvisation A (CD Track #5), 65–69
Trio Improvisation B (CD Track #6), 71–73
Trio Improvisation C (CD Track #7), 71–73
Trio Improvisation D (CD Track #8), 75–80
Trio Improvisation E (CD Track #9), 75–80
Trio Improvisation F (CD Track #10), 81–84
Trio Improvisation G (CD Track #11), 81–84

U

University of Missouri–Kansas City Conservatory of Music sessions
  Duet Improvisation, 61–63
  Expanding Quartet Improvisation, 85–89
  overview of, 59–60
  Trio Improvisation A, 65–69
  Trio Improvisation B, 71–73
  Trio Improvisation C, 71–73
  Trio Improvisation D, 75–80
  Trio Improvisation E, 75–80
  Trio Improvisation F, 81–84
  Trio Improvisation G, 81–84

V

Verb music, 28
Vibraphone, 5, 15
Vocabulary building, 51–52
Vocal teachers, 14

W

Wang, Huei-hsien, 14
Winter, Paul, 4
Woodruff, Paul, 41
Workshop, 34
Wyre, John, 1

X

Xylophones, 16

LIMITED WARRANTY

Taylor & Francis Group warrants the physical disk(s) enclosed herein to be free of defects in materials and workmanship for a period of thirty days from the date of purchase. If within the warranty period Taylor & Francis Group receives written notification of defects in materials or workmanship, and such notification is determined by Taylor & Francis Group to be correct, Taylor & Francis Group will replace the defective disk(s).

The entire and exclusive liability and remedy for breach of this Limited Warranty shall be limited to replacement of defective disk(s) and shall not include or extend to any claim for or right to cover any other damages, including but not limited to, loss of profit, data, or use of the software, or special, incidental, or consequential damages or other similar claims, even if Taylor & Francis Group has been specifically advised of the possibility of such damages. In no event will the liability of Taylor & Francis Group for any damages to you or any other person ever exceed the lower suggested list price or actual price paid for the software, regardless of any form of the claim.

Taylor & Francis Group specifically disclaims all other warranties, express or implied, including but not limited to, any implied warranty of merchantability or fitness for a particular purpose. Specifically, Taylor & Francis Group makes no representation or warranty that the software is fit for any particular purpose and any implied warranty of merchantability is limited to the thirty-day duration of the Limited Warranty covering the physical disk(s) only (and not the software) and is otherwise expressly and specifically disclaimed.

Since some states do not allow the exclusion of incidental or consequential damages, or the limitation on how long an implied warranty lasts, some of the above may not apply to you.